ESSENTIALS OF
HUMAN PHYSIOLOGY

PHILLIP SHEELER
Professor of Biology
California State University, Northridge

CENTRISEP PUBLISHING
NORTHRIDGE, CA

ESSENTIALS OF
HUMAN PHYSIOLOGY

Cover: The photograph on the cover of this book is a transmission electron-photomicrograph of a thin section through striated muscle tissue. The author thanks Richard Chao for his permission to use this photograph.

To my wife, Annette

ESSENTIALS OF
HUMAN PHYSIOLOGY

TABLE OF CONTENTS

CHAPTER 7

THE CIRCULATORY SYSTEM 101

CHAPTER 8

THE BLOOD 119

CHAPTER 9

THE IMMUNE SYSTEM 135

CHAPTER 10

THE RESPIRATORY SYSTEM 153

CHAPTER 11

THE DIGESTIVE SYSTEM 167

CHAPTER 12

THE EXCRETORY SYSTEM 181

CHAPTER 13

THE ENDOCRINE SYSTEM 191

CHAPTER 14

REPRODUCTION AND INHERITANCE 211

PREFACE

I have taught the Human Physiology course at California State University, Northridge for 23 years. At our institution, the course is designed for students majoring in such diverse fields as health science, nutrition science, physical education, physical therapy, pre-nursing studies, psychology, and a number of related disciplines. The human physiology course is not designed for students majoring in biology, although a number of biology majors do elect to take the course, as do a number of students whose majors are not health-related (e.g., art, history, etc.). Classes are filled, therefore, with students of diverse backgrounds and goals.

Over the years, I've used many different textbooks for the course, most of them quite good; indeed, as long as they are up-to-date in their presentations, nearly all modern textbooks of human physiology can adequately serve a course such as this one. In recent years, however, there has been a trend in human physiology texts toward greater and greater amplification of subject matter. As a result, textbooks in this area have become unnecessarily elaborate (and expensive for the students), and their coverage has been unduly extended. The result is textbooks whose coverage is so extensive that only a small fraction of the book's content can be covered in a one-semester course. Many students are intimidated or overwhelmed by such a text's scope and coverage and are deterred from doing the necessary reading.

Because of this, and at the urging of many of my students, I decided to prepare my own textbook, the scope and content of which more realistically interface with the coverage of the course. I have written and illustrated the book using "desk-top publishing" methods, so that the book's production cost could be reduced. Accordingly, this book covers what I consider to be the essentials of a one-semester (about 40 lecture hours) human physiology course. I have concentrated on the more important and interesting topics and I have tried to present the material in a clear, concise, and readable manner.

For those of you who are interested in such things, this book was written using various models of the Macintosh[1] personal computer running Microsoft's "Word[2]" word-processing program. All of the artwork was done using either Apple's "MacDraw[1]" program or using Silicon Beach Software's "SuperPaint[3]." Camera-ready copy of the text and graphics was prepared using either an Apple Laserwriter[1] Plus or Laserwriter IISC laser printer.

Phillip Sheeler
Northridge, California
April, 1989

---

1 Macintosh, MacDraw, and Laserwriter are trademarks of Apple Computer Inc.
2 Word is a trademark of Microsoft Corporation.
3 SuperPaint is a trademark of Silicon Beach Software.

INTRODUCTION

SOME BASIC CONCEPTS

Physiology is the branch of **biological science** that attempts to explain in chemical, physical, and molecular terms the multitude of phenomena displayed by living things. Physiology has several sub-branches of its own, including *animal* physiology, *plant* physiology, and *microbial* physiology. This book is concerned with a specific aspect of animal physiology, namely **human physiology**–the study of how the human body works.

LEVELS OF ORGANIZATION OF THE BODY

The human body is an extremely complex structure. To simplify its study, it is necessary to subdivide the body into a number of different functional parts, each of which can then be considered separately. The major functional subdivisions of the human body are its **organ-systems**, each organ-system having a rather specialized function of its own. The major organ-systems of the body are:

1. the **muscle** system
2. the **nervous** system
3. the **receptor** system
4. the **cardiovascular** system
5. the **immune** system
6. the **respiratory** system
7. the **digestive** system
8. the **excretory** system
9. the **endocrine** system
10. the **reproductive** system

Each organ-system also has a hierarchy of structural and functional parts. Organ-systems are comprised of a number of **organs**. For example, the eyes and the ears are organs of the receptor system; the stomach, pancreas, and small intestine are organs of the digestive system; and the kidneys and urinary bladder are organs of the excretory system. Each organ is formed by an assemblage of **tissues**. For example, the stomach contains *muscle* tissue, *epithelial* tissue, *connective* tissue, *vascular* tissue, and *nerve* tissue. Each tissue is made up of large numbers of individual **cells**, the cells of each tissue sharing properties common to the cells of other tissues and also possessing unique properties (e.g., muscle cells *contract*, nerve cells *conduct*, endocrine cells *secrete*, and so on). Even cells may be subdivided into distinct structural and functional components; these are the sub-cellular **organelles** (e.g., *nucleus, mitochondria, ribosomes*, etc.) and their respective molecular and atomic constituents.

Thus, in order of decreasing scope and increasing functional specificity, the levels of organization of the human body are:

1. **organ-systems**
2. **organs**
3. **tissues**
4. **cells**
5. **organelles**
6. **molecules**

Most of the remaining chapters in this book deal with the organization and functions of the tissues and organs that make up the body's organ-systems. The balance of this first chapter is devoted to a review of some basic chemical concepts that are fundamental to your understanding of the material that appears in succeeding chapters. If you are already familiar with these concepts, skip to Chapter 2.

ELEMENTS AND THE STRUCTURE OF ATOMS

Elements are fundamental units of matter and include, for example, *carbon*, *hydrogen*, *oxygen*, and *nitrogen*–the most abundant of the elements of human cells and tissues. Although more than 100 different elements occur in nature, only a small number of these are found to any appreciable extent in the cells and tissues of the human body. Elements are said to be "fundamental" in that they cannot be changed into other elements as a result of conventional chemical reactions. To be sure, there are instances in which one element is changed into another as the result of *radioactive decay*; however, such changes involve the atomic nucleus and do not fall within the realm of chemical reactions that characterize living things. The chemical reactions that characterize living things involve changes in the electrons surrounding the nucleus of an atom and do not involve intra-nuclear rearrangements.

An **atom** is the basic form of an element and is, in turn, comprised of *elementary particles*. The atoms of all known elements are formed from the same elementary particles and differ only in the numbers of these particles. Insofar as human physiology is concerned, three kinds of elementary particles take on importance; these are **protons**, **neutrons**, and **electrons**, and it is the electrons that participate in the chemical reactions that characterize each cell's activities.

Of the three types of elementary particles, two of them–the protons and neutrons–are found in the "core" or **nucleus** of the atom, whereas the electrons are found in a surrounding sphere (for convenience, they may be thought of as orbiting the nucleus, much as the planets orbit the sun; Fig. 1-1). The protons possess *positive* electrical charge, whereas the neutrons are electrically *neutral*; thus an atomic nucleus carries positive charge. In contrast, the electrons that surround the nucleus are *negatively* charged. Since like charges repel one-another and opposite charges attract one-another, there is a force attracting the negatively-charged electrons of an atom to its positively-charged nucleus.

In a *complete* atom, the sphere of electrons that surrounds the nucleus contains the same number of negative charges as the nucleus contains positive charges. Therefore, the atom as a whole is electrically neutral. Since the nucleus' neutrons are electrically neutral, this also implies that the number of protons in the nucleus must equal the number of surrounding electrons. To gain some insight into the relative size of the atom and the space occupied by its parts, if the nucleus of a carbon atom were magnified to the size of a baseball, then the outermost electrons would be about 100 yards away (about the length of a football field). Thus, an atom is largely empty space.

The electrons of an atom occur in layers or **shells** at specific distances from the atomic nucleus (Fig. 1-2). Each shell can accommodate a certain number of electrons. The innermost shell can contain up to 2 electrons, the next shell no more than 8, and succeeding shells contain either the same or a greater number of electrons than the shell before (i.e., 2, 8,

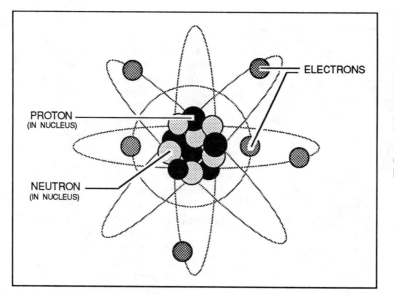

Figure 1-1
Structure of an **atom**. The **nucleus** of an atom contains **protons** and **neutrons**; the protons are positively-charged, whereas the neutrons carry no charge (i.e., they are "neutral"). Around the nucleus orbit a number of **electrons**, each electron bearing a negative charge.

8, 18, 32, etc.). Larger and larger atoms contain correspondingly greater numbers of electrons, and generally the inner shells fill before the outer shells. The chemical "reactivity" of an atom is based upon the extent to which its outermost shells are filled.

MOLECULES AND CHEMICAL BONDS

Two or more atoms may be linked together to form a **molecule**, with the linkages between the atoms referred to as **chemical bonds**. Molecules that are comprised of more than one kind of atom are called **compounds**. Water, the most abundant molecular substance in the human body, is a compound that is formed from one atom of oxygen and two atoms of hydrogen.

Each type of atom can combine with a specific *maximum* number of other atoms. For example, a chlorine atom can combine with *one* other atom, an oxygen atom can combine with as many as *two* atoms, and a carbon atom can combine with as many

as *four* atoms. Such observations lead to the concept of **valency**, according to which a particular atom is thought to possess a fixed number of bonding sites that could combine with the corresponding bonding sites of other atoms. Thus, from the information given above, we would say that chlorine exhibits a valence of *one*, oxygen a valence of *two*, and carbon a valence of *four*. Chemists explain valency in terms of the properties of the outer electrons (i.e., the so-called *valence electrons*). Elements whose outermost shell contains only one electron (or whose outermost shell is one electron short of being filled) have a valence of one. The elements sodium and potassium have only one electron in their outermost shell, and therefore their valence is one. Likewise, elements with a valence of two have an outer shell that is either two electrons short or has only two electrons in it. Carbon, with its valence of four, has four electrons in its second shell; thus, carbon can be thought of as having *either* a four-electron deficiency *or* a four-electron surplus. This special characteristic of carbon is one of the properties that makes carbon atoms so

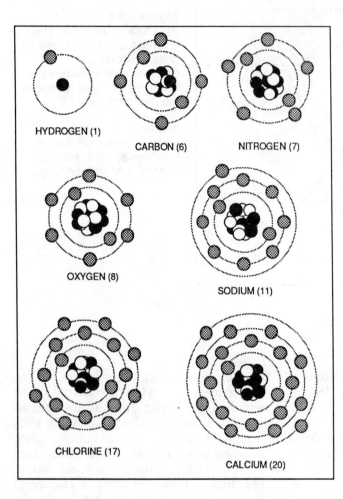

HYDROGEN (1)

CARBON (6)   NITROGEN (7)

OXYGEN (8)

SODIUM (11)

CHLORINE (17)

CALCIUM (20)

Figure 1-2
In a complete atom, there are as many electrons orbiting the atom's nucleus as there are protons in the nucleus. The orbiting electrons occur in a series of encapsulating shells: up to 2 electrons in the first shell, up to 8 electrons in the second shell, and so on. In the diagram, the number in parentheses after the atom's name is the element's **atomic number**. This number corresponds to the number of protons in the nucleus, and in a complete atoms is also the number of orbiting electrons. Thus, calcium (atomic number 20) has 20 protons in its nucleus and has 20 orbiting electrons: 2 in the first shell, 8 in the second shell, 8 in the third shell, and 2 in the fourth shell.

important in the chemistry of living things.

Formation of **bonds** between atoms can be thought of as a tendency of the atoms to avoid partially-filled outer shells. Atoms can fill their outermost shells in two ways: (1) by **ionic bonding** or (2) by **covalent bonding**. Ionic bonding involves the transfer of electrons from an atom with an outer shell surplus to an atom with an outer shell deficiency. Covalent bonding involves the *sharing* of outer shell electrons between two atoms.

Ionic Bonds and Ions

Ionic bonds are formed by the transfer of one or more outer-shell electrons from one atom to one or more other atoms. Because the transferred electrons are negative charges, the formation of an ionic bond also results in a charge-transfer between the atoms. Atoms that receive electrons in such a transfer become more negative, while atoms donating electrons become more positive (or less negative). The bond that holds the two atoms together is the attractive force that exists between the two opposite charges. A common example of an ionic bond is the bond between sodium and chlorine in sodium chloride (NaCl) crystals (ordinary table salt; Fig. 1-3). The sodium atoms in the salt crystal give up their single outer shell electron to chlorine, thereby becoming positively charged. By accepting these

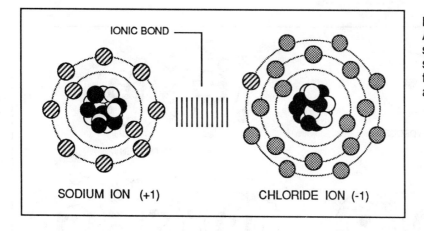

Figure 1-3
An **ionic bond**. In sodium chloride crystals, sodium and chloride ions form ionic bonds with one another.

electrons, the chlorine atoms fill their outermost shell and take on a negative charge.

Atoms that have gained or lost electrons are called **ions**. A *positively-charged* ion is called a **cation**, and a *negatively-charged* ion is called an **anion**. In sodium chloride, the sodium is the cation (represented by the symbol Na^+) and the chloride is the anion (represented by Cl^-). Ions can also be composed of more than one atom. In the salt *ammonium chloride* (NH_4Cl), the ammonium portion which is composed of one nitrogen and four hydrogens (NH_4^+) becomes a cation when the atoms give up an electron to chlorine (Cl^-).

As a second illustration, consider the case of the salt *calcium chloride* ($CaCl_2$) in which *two* electrons are transferred. Thus, calcium forms *two* ionic bonds–one with each of the two chlorines. Calcium has only two electrons in its outermost shell; one electron is donated to each of the chlorines. Since two electrons are donated, the calcium cation carries two positive charges and is represented by the symbol Ca^{++}.

Covalent Bonds

Another way to fill the outermost electron shell is through the formation of **covalent** **bond**s. Covalent bonds are not formed by the transfer of electrons, but through the *sharing* of electrons. A simple example of covalent bonding occurs in molecular hydrogen or H_2 (Fig. 1-4). In water (H_2O), oxygen and two hydrogen atoms share outer electrons. A covalent bond that results from the sharing of one pair of electrons between two atoms is called a **single bond**.

In a number of compounds, certain of the atoms share *more than one* pair of electrons in order to complete their outermost shells. Bonds formed by sharing more than one electron pair are called **multiple bonds**. For example, a **double bond** is one in which *two* pairs of electrons are shared; in a **triple bond**, *three* pairs of electrons are shared.

Gaseous oxygen occurs in the form of diatomic molecular oxygen. An oxygen atom has 6 outer shell electrons; therefore, two more would complete its shell. In molecular oxygen, this occurs by sharing two pairs of electrons (Fig. 1-4). Thus, the oxygen atoms are linked by a double bond. The shorthand for a double bond is two parallel dashes, so that the oxygen molecule may be depicted O=O. Another example of a molecule that contains double bonds is carbon dioxide (CO_2). As we already have seen, carbon's outer shell contains four electrons but has a capacity for eight. Therefore, the outer shells of the carbon atom and the two oxygens are completed when the carbon and oxygen

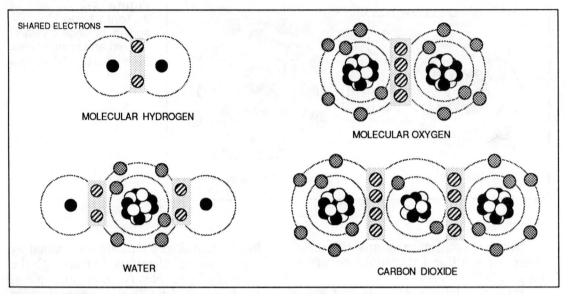

Figure 1-4
Covalent bonds. Covalent bonds are formed when two or more atoms *share* electrons, such that each atom's outer shell is filled. In molecular hydrogen, two hydrogen atoms share their single electron. When two electrons are shared (as shown here for molecular hydrogen and water), the covalent bond is called a **single bond**. When four electrons are shared (as shown here for molecular oxygen and carbon dioxide), the covalent bond is called a **double bond**.

share two pairs of electrons (Fig. 1-4).

MOLECULAR WEIGHT

Just as atomic weight is the sum of the weights of the elementary particles of an atom, **molecular weight** is the sum of the weights of the atoms that form a molecule. The units of molecular weight are *daltons*. For example, water's molecular weight is 18 daltons, which is the sum of the atomic weights of oxygen (16 daltons) and the two hydrogen atoms (one dalton each). The molecular weight of carbon dioxide (CO_2) is 44 daltons ($12 + [2 \times 16]$), and the molecular weight of ammonia (NH_3) is 17 daltons ($14 + [3 \times 1]$).

POLAR COVALENT BONDS

When two identical atoms form a covalent bond, the electrons are shared equally between the two atomic nuclei. When different atoms form a covalent bond, the electrons are not shared equally. Instead, one of the two atoms attracts the shared electrons more strongly than does the other atom. The attraction of an atom for electrons is called **electrophilia**. In the N–H bond, the nitrogen atom is more electrophilic than the hydrogen atom; in the O–H bond, it is the oxygen that is the more electrophilic. Electrophilia is based on the number of protons in the atom's nucleus. Thus nitrogen, with its seven protons, is more electrophilic than hydrogen which has only one proton. Since shared electrons may be more strongly attracted to the nucleus of one atom than the other, the bond that they create becomes **polar**; that is, at one end of the bond the electrical charge is slightly different than at the other end of the bond.

The bonds between oxygen and hydrogen in water are examples of **polar covalent bonds**. In a water molecule, the shared electrons are more strongly attracted to the oxygen atom than they are to the hydrogen atoms. As a result, the shared electrons spend more time near the oxygen nucleus than they do near each hydrogen atom. Each hydrogen atom is therefore said to possess a *partial positive charge* (symbolized by ∂^+), and the oxygen atom is said to have a *partial negative charge* (that is, ∂^-).

Hydrogen Bonds

Because the hydrogen nucleus is so much less electrophilic than the nuclei of carbon, oxygen, and nitrogen atoms, the covalent bonds that hydrogen forms with these atoms are polar. The result is that the hydrogen carries a partial positive charge. Partially positive hydrogen atoms that are part of one molecule may be attracted to partially negatively charged atoms in other molecules. When this occurs a weak bond is created between the two molecules. Such a bond is called a **hydrogen bond**, because the two molecules are linked together by the hydrogen nucleus (Fig. 1-5). Although they are individually weak, the sheer numbers of hydrogen bonds create important stabilizing forces between biological molecules. Hydrogen bonds can be formed *within* as well as *between* molecules. Many of the very large molecules that are found in the body (e.g., proteins) owe their precise shape to the many hydrogen bonds that these molecules contain.

WATER AND SPHERES OF HYDRATION

Water is the most abundant molecular constituent of the body and in most tissues cellular water often exceeds 80% of the total cell weight. Many of the physical and chemical events that take place in cells are the results of reactions between compounds that are dissolved in the cellular water. Since many of these reactions depend upon the special chemical properties of the surrounding water molecules, it is apparent that the role of water is an active one. Many of the special properties of water stem from the polar nature of the bonds between its oxygen and hydrogen atoms.

Of special biological significance is the tendency of water molecules to arrange themselves around dissolved ions such as Na^+ and K^+ thereby forming **spheres of hydration** (Fig. 1-6). For example, sodium ions form hydration spheres containing several water molecules. The at-

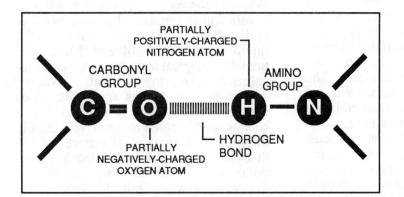

Figure 1-5
Hydrogen bonds. Partially positively-charged hydrogen atoms that are a part of one molecule may be attracted to partially negatively-charged atoms in other molecules. The resulting bond created between the two molecules is called a *hydrogen bond*.

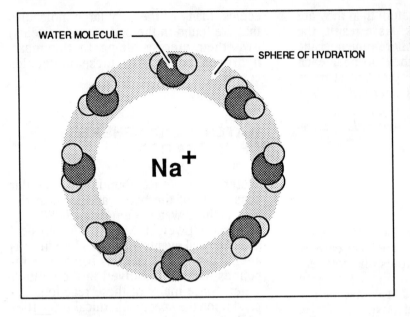

WATER MOLECULE

SPHERE OF HYDRATION

Na$^+$

Figure 1-6.
Spheres of hydration.
Water molecules are attracted to anions and cations and form enclosing spheres of hydration. Shown here is a sphere of hydration formed around a sodium ion.

tractive force between the positively-charged sodium ion and the water involves the partial negative charges of the oxygen atoms of the water molecules. The one or more spheres of hydration that surround an ion give the ion some unexpected chemical and biological properties. For example, for many years physiologists were puzzled by the observation that cations of high atomic weight were able to pass through the pores of membranes more quickly than cations of lower atomic weight (i.e., the smaller ions were expected to pass through more readily). The larger number of water molecules that form hydration spheres around a small ion than a large ion turned out to be the explanation. That is, the hydration sphere around a small ion causes the ion to assume a much larger *effective size*. The reason that smaller ions attract a larger number of water molecules than do larger ions is due to the differences in the numbers of electrons around the ion's nucleus. A large monovalent cation possesses the same electrical charge as a small monovalent cation, but the number of electrons in the shells surrounding the nucleus is greater in the larger cation. The greater number of electrons present acts to reduce the interaction of the ion's positive nucleus

with the partially-negative portions of water molecules. The differences in effective sizes of hydrated *divalent* cations, *trivalent* cations, and so on may be explained in the same way. Spheres of hydration are also formed around anions such as Cl$^-$. In such instances, the attractive force occurs between the anion and the partially-positive hydrogen atoms of the water molecules.

SOLUTIONS AND THE CONCEPT OF CONCENTRATION

When a substance is dissolved in a liquid, a **solution** is formed. The substance that is dissolved is called the **solute** and the liquid is called the **solvent**. Most solutions of biological importance are *aqueous* solutions (i.e., the solvent is water). Although most solutes are solids, gases can also be dissolved in a liquid. When solutes are dissolved in water, it is usually through the formation of spheres of hydration around the solute molecules. Nonpolar substances such as fats do not dissolve in water but do dissolve in *organic* solvents, such as acetone.

It is important to be able to specify how much solute is dissolved in a given volume of solvent—that is, the *concentra-*

tion of of the solution. Although concentration can be expressed in a number of different ways, each method reflects the *number* of solute molecules per unit volume of solution. This is because, other things being equal, twice as many molecules dissolved in a given volume of solvent will be twice as effective in doing whatever the molecules do. For example, an oxygen-transporting blood pigment like *hemoglobin* found in one organism may be five times the size of one found in another organism, but if each hemoglobin molecule can carry only one molecule of oxygen, then for any given quantity of oxygen to be carried, five times as much of the larger of the two hemoglobins would be required.

Most physiologists use **molarity** to express concentration. Recall that atoms of each element have a characteristic atomic weight and that a compound has a characteristic molecular weight. Yet another important measure of weight is the **gram atom**, which is defined as the number of grams of an element that is numerically equal to the element's atomic weight. A gram atom of carbon weighs 12 grams and a gram atom of magnesium weighs 24 grams. However, although a gram atom of magnesium weighs twice as much as a gram atom of carbon *both* contain *the same number* of atoms. The number of atoms present in one gram atom of any element is a *constant* called **Avogadro's Number** and is equal to 6.023×10^{23}.

Likewise, a **gram molecule** of a compound is a number of grams numerically equal to the compound's molecular weight. For example, the atomic weight of oxygen is 16; therefore a gram molecule of oxygen (O_2) weighs 32 grams. Similarly, the molecular weight of water is 18, and therefore a gram molecule of water weighs 18 grams. Just as a gram atom of an element contains Avogadro's Number of atoms, a gram molecule of a compound contains Avogadro's Number of molecules.

A quantity of any element or compound that contains Avogadro's Number of atoms or molecules is also said to represent one **mole** of that substance. Thus, a mole of water weighs 18 grams and contains 6.023×10^{23} molecules. Although the term mole infers the same quantity of a substance as gram atom or gram molecule, the mole is the more frequently used expression of quantity, since it applies to *both* individual elements and compounds.

A solution that has a molarity of 1.0 contains one mole of a solute per liter of solution. Put another way, one liter of a 1.0 molar solution contains 6.023×10^{23} molecules (or atoms) of the solute. The molarity of a solution is symbolized by the letter **M**; thus a one molar solution of the sugar *sucrose* would be referred to as 1.0 **M** sucrose; one-tenth molar as 0.1 **M**, and so on.

The concentrations of most solutes in the cells and tissues of living things are very much lower than one molar. Therefore, it becomes more practical to express concentration in units that are correspondingly smaller such as *millimoles* (thousandths of a mole), *micromoles* (millionths of a mole), or even *nanomoles* (billionths of a mole). For example, a one millimolar solution (abbreviated 1 m**M**) contains one-thousandth of a mole of solute per liter of solution. In the same way that a millimole is one-thousandth of a mole, a *milligram* (mg) is one-thousandth of a gram. Thus, a 0.5 m**M** glucose solution contains 90 milligrams per liter.

DISSOCIATION AND THE CONCEPT OF pH

Many substances dissociate into ions when they are dissolved in water. These compounds are called **electrolytes** because they are able to conduct electricity through the solution. Compounds that do not dissociate into ions are called **non-electrolytes** and are poor conductors of electricity. Electrolytes may be divided into three classes; these are (1) **acids**, (2) **bases**, and (3) **salts**.

Acids and bases can be defined in terms of their effects on the concentrations of hydrogen ions (H+) and hydroxyl ions (OH-) in an aqueous solution. Even pure water contains some H+ and OH- because water itself undergoes a small amount of dissociation. The dissociation of water molecules in pure water produces H+ and OH- concentrations that have a molarity of only 0.0000001 (i.e., 0.1 micromolar). An **acid** *increases* the concentration of H+ when added to a solution (or *decreases* the concentration of OH-); whereas a **base** *increases* the concentration of OH- (or *decreases* the concentration of H+). For example, *hydrochloric acid* (HCl) dissociates in water to form H+ and Cl-, thereby raising the concentration of H+. On the other hand, *sodium hydroxide* (NaOH) is a base because it dissociates into Na+ and OH-, thereby raising the hydroxyl ion concentration of a solution (Fig. 1-7, top).

When salts such as NaCl and KCl dissociate in water, neither a hydrogen ion nor a hydroxyl ion is produced and the amounts of hydrogen ions and hydroxyl ions in the water remain unchanged. Mixing an acid with a base in the right proportions produces water and salt ions, as occurs when sodium hydroxide is mixed with hydrochloric acid (Fig. 1-7, bottom).

Hydrochloric acid is said to be a "strong" acid because it undergoes a high degree of dissociation when it is dissolved in water. When HCl is dissolved in water,

nearly all of the HCl molecules dissociate into ions, and only a few HCl molecules remain intact. In contrast, *acetic acid* dissociates much less completely when dissolved in water. When acetic acid is added to water, *acetate* ions and H+, are produced, but substantial amounts of undissociated acetic acid remain. Acetic acid is therefore considered a "weak" acid because of its small degree of dissociation. In a like manner, there are also *weak bases* and *weak salts*.

Most of the chemical reactions that take place in the body are especially sensitive to the concentration of hydrogen ions present. For example, the digestion of food proteins inside the stomach requires a highly acidic environment; in contrast, the opposite is true in the small intestine, where digestion of the food requires a basic environment. The internal fluid of most cells is neither acidic nor basic; rather it is said to be *neutral*, since the concentrations of H+ and OH- are about the same as in pure water (namely, 1.0×10^{-7} **M**). Small variations in the H+ or OH- concentrations of the intracellular fluid prevent a cell from functioning properly and may even be fatal to the cell.

The term **pH** is used to indicate how acidic or how basic a solution is. The pH value of a solution is defined by the equation $\mathbf{pH = -log_{10} [H^+]}$. Thus, the pH of pure water, in which $[H^+] = 1.0 \times 10^{-7}$

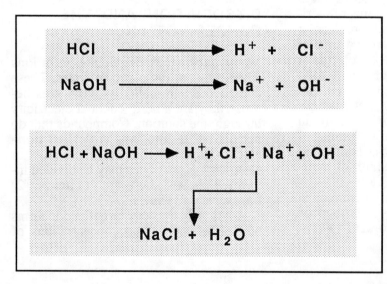

Figure 1-7

Dissociation of acids and bases.

M, is $-\log_{10} (10^{-7}) = 7$. Water is said to have a neutral pH since the concentrations of H^+ and OH^- are equal. Thus, on the pH scale, 7.0 represents neutrality. Solutions that have pH values less than seven are said to be "acidic" whereas those that have a pH greater than seven are "basic" (or "alkaline"). Because pH is a logarithmic function, one unit of pH represents a 10-fold difference in hydrogen ion concentration. A solution having a pH of 6 has ten times the hydrogen ion concentration as one of pH 7. A solution of pH 5 has 100 times the hydrogen ion concentration as one of pH 7, and so on.

Since most intracellular processes operate most effectively when the pH is near 7.0, **buffers** are used in order to maintain a pH near neutrality. The body's natural buffers are mixtures of weak acids and their anions and are able to neutralize both acidic and basic solutions. A number of different buffer mixtures are present in cells, in the bloodstream, and in other biological fluids. An example of an important buffer of the bloodstream is the weak acid *carbonic acid* (H_2CO_3) and its anion, *bicarbonate* (HCO_3^-). The role of the buffer is to combine with hydrogen (or hydroxyl) ions that may be added to the solution, so that the solution's pH does not change.

A weak acid undergoes only partial dissociation, so that a solution of the weak acid contains a mixture of intact weak acid molecules, hydrogen ions, and anions having a characteristic and constant numerical ratio. Therefore, if some H^+ is added to the solution, this disturbs the equilibrium balance by raising the $[H^+]$ value. To restore the equilibrium ratio, some H^+ reacts with the anion to form more of the weak acid. As a result, the added H^+ disappears. Similarly, addition of OH^- to the buffer solution results in the conversion of some of the H^+ already present into water (remember that water itself undergoes a small amount of dissociation into H^+ and OH^-). The loss of H^+ is "made up" by the dissociation of a little more of the weak acid, so that in the end the solution's pH is maintained.

With these basic chemical concepts in mind, let us now move on to a consideration of the body's chemistry.

CHEMISTRY OF THE BODY

Much of the body's weight is represented by water. For example, in the average young adult male, water accounts for more than 60 percent of the body weight; in young adult females, water accounts for about 50 percent of the total body weight. The remaining molecules can be divided into two broad categories: **organic** and **inorganic**.

Generally speaking, organic substances are compounds peculiarly associated with living things, whereas inorganic compounds are found both in living things and as regular constituents of the inanimate world. Biochemists employ a much stricter definition of an organic substance, limiting this category to compounds that contain both *carbon* and *hydrogen* atoms. There are some substances (e.g., carbon dioxide) that do not properly fit the biochemical definition of "organic" but which clearly are associated with the activities of living things. Inorganic substances include such things as salts and minerals and typically are compounds of relatively low molecular weight. In this chapter, our principal concern will be with the structure and chemistry of organic compounds.

In most organic compounds of the body, carbon forms a backbone for the molecule, as for example, in the common body sugar called **glucose** (Fig. 2-1). Other atoms (or groups of atoms) are attached to the carbon backbone, and usu-ally these give the compound its unique chemical and physiological properties. Because organic compounds frequently contain several carbon atoms, each of the carbon atom is assigned a number so that it (or the group of atoms to which the carbon is attached) can be specifically identified.

Many biologically important compounds form *ringed* structures. For example, the six carbon atoms of *benzene* (Fig. 2-1) form a closed loop in which the carbon atoms are linked by alternating single and double bonds. The remaining carbon bonds are formed with hydrogen. For convenience and simplicity, ringed compounds like benzene are often drawn as geometric figures, in which it is understood that each corner of the figure is occupied by a carbon atom or CH-group. Even in this abbreviated form, the locations of the double bonds are identified. In some ringed structures, nitrogen or oxygen atoms are members of the ring; the ringed forms of the sugars glucose and ribose are common examples. The positions of the non-carbon members of a ring are always noted by using the element's symbol (Fig. 2-1). Nearly all of the major ringed structures of importance in the body contain 5 or 6 members. Compounds that contain ringed structures are usually called *aromatic* compounds because of their characteristic aromas or odors.

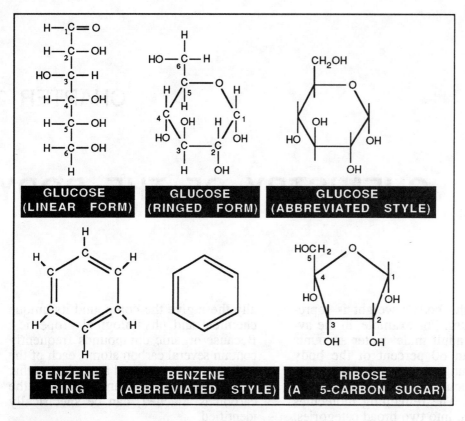

Figure 2-1
Formulas of glucose, benzene, and ribose. Glucose exists in *linear* and *ringed* forms. In the abbreviated formulas, individual carbons of a ring are not shown (they occur at each vertex). When a bond appears to have no attached group, this implies that the attached group is a hydrogen atom.

MAJOR GROUPS OF ORGANIC COMPOUNDS

For convenience, we can assign all of the common organic compounds of the body into five major groups: (1) **organic acids**, (2) **amino acids**, (3) **sugars**, (4) **nucleotides**, and (5) **lipids**.

Organic Acids

Organic acids are compounds that contain one or more **carboxyl** (i.e., COOH) groups. One of the simplest of these compounds is *acetic acid* (Fig. 2-2). A number of important organic acids have two carboxyl groups. An example is the four-carbon compound *malic acid*, which also has a **hydroxyl** group attached to the second carbon atom (Fig. 2-2). Acids that contain a carboxyl group dissociate incompletely, and they are therefore considered weak acids.

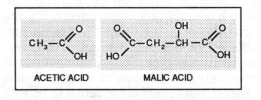
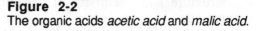

Figure 2-2
The organic acids *acetic acid* and *malic acid*.

Amino Acids

Amino acids are especially important organic compounds because they are the

14

building blocks of **proteins**. Amino acids are organic acids that contain a basic **amino** group in addition to a carboxyl group. In this regard, amino acids are somewhat unusual since they contain *both* acidic and basic groups. The general formula of an amino acid is shown in Fig. 2-3; in the general formula, the letter **R** represents one of a number of different chemical groups that vary from a single hydrogen atom to more complex chemical structures. Thus, it is the different R groups that give rise to the different kinds of amino acids (Fig. 2-4). The R groups are also known as the amino acids' **side-chains**.

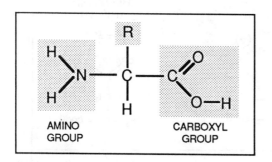

AMINO GROUP

CARBOXYL GROUP

Figure 2-3
General formula of an amino acid.

The tissues of the body contain many large molecules called **macromolecules** which consist of long chains of smaller subunits. Among the most important macromolecules are the **proteins** formed by combinations of amino acids. The 20 amino acids most commonly found in proteins are shown in Figure 2-4. The typical protein contains hundreds of amino acids that form one or more chains called **polypeptides**. The amino acids of a polypeptide are linked together by **peptide bonds** formed between the amino group of one amino acid and the carboxyl group of its neighbor (Fig. 2-5). When two amino acids are linked together by a peptide bond, a **dipeptide** is produced; the linking of three amino acids by peptide bonds forms a **tripeptide**, and so on.

Sugars

Sugars are a diverse collection of compounds in which the carbon backbone of the molecule forms bonds with several hydroxyl groups and contains at least one **carbonyl** (i.e., C=O) group. In nearly all common sugars, the numbers of oxygen and hydrogen atoms are such that the sugar may be abbreviated $C_n(H_2O)_n$. That is, for every carbon atom in the molecule there is the equivalent of one molecule of water. For most of the common sugars, the value of n is 5 or 6. When 5 carbons are present, the sugar is a **pentose**; 6 carbon sugars are called **hexoses**. The chemical structures of the hexose *glucose* and the pentose *ribose* are shown in Fig. 2-1.

The most important common sugar is **glucose** (also called **dextrose**). As seen in Fig. 2-1, glucose (as well as many other sugars) can exist in an "open" (i.e., linear chain) form or as a ringed structure; in cells and tissues, it is the ringed forms of the sugars that predominate. Through what are known as *condensation* reactions, one molecule of glucose (or other sugar) can be linked to a second sugar molecule, producing what is called a **disaccharide** (Fig. 2-6). Each sugar of the disaccharide is called a **monosaccharide** and the two monosaccharides are linked by a **glycosidic** bond. One of the most abundant disaccharides is **sucrose** (i.e., ordinary "table sugar"), which is formed from one molecule of glucose and one molecule of **fructose**. Sucrose is the sugar used to sweeten drinks (coffee, tea, etc.). Another common disaccharide is **maltose**, which is formed from two glucose molecules. Macromolecules called **polysaccharides** are formed by linking large numbers of sugars together. **Glycogen** is the major polysaccharide of the human body and represents the form in which the body stores sugar.

Nucleotides

Nucleotides are the subunits (building blocks) of a family of important macro-

15

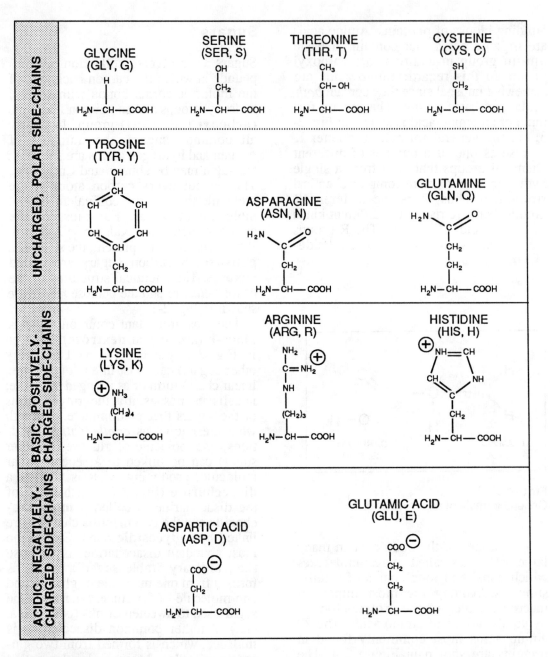

Figure 2-4
The 20 different amino acids that regularly occur in proteins. It is the varying chemical composition of the R groups or **side-chains** that distinguish one amino acid from another. The conventional three-letter and one-letter abbreviations used for the amino acids are also shown.

molecules called **nucleic acids**. The nucleic acids are information-carrying molecules, which among other things make up the units of heredity called **genes**. A nucleotide consists of three different chemical groups: (1) a **pentose**, (2) a **phosphate** group, and (3) one of several different ringed compounds called

16

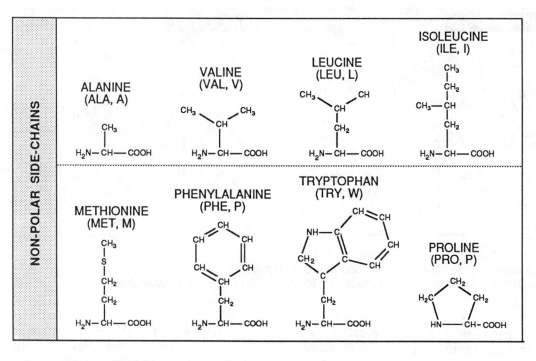

Figure 2-4 *(continued).*

nitrogenous bases (Fig. 2-7). In nucleic acids, the nucleotides are strung together to form long chains or **polynucleotides**. The backbone of the chain consists of alternating pentose and phosphate groups. The nitrogenous bases are linked to each of the sugars but project away from the molecule's backbone (Fig.

2-8). The pentose of a nucleotide is either **ribose** (Fig. 2-1) or **deoxyribose** (in which the hydroxyl group attached to carbon atom number 2 is replaced by a hydrogen atom; see Fig. 2-1). Nucleic acids in which the sugar is ribose are called **ribonucleic acids** (usually abbreviated **RNA**), whereas nucleic acids in which

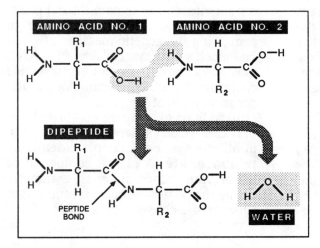

Figure 2-5
Formation of a **dipeptide** by linking two amino acids. The bond linking the amino acids (a **peptide bond**) is formed between the nitrogen atom of the amino group of one amino acid and the carbon atom of the carboxyl group of the other amino acid.

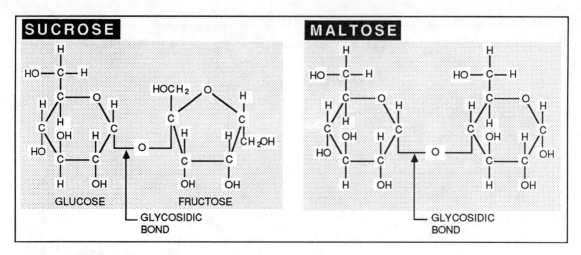

Figure 2-6
The disaccharides **sucrose** (left) and **maltose** (right). Sucrose is formed from glucose and fructose, whereas maltose is formed from two glucose molecules.

the sugar is deoxyribose are called **deoxyribonucleic acids** (abbreviated **DNA**).

The nitrogenous bases of nucleotides may be either **pyrimidines** or **purines**. As seen in Figure 2-7, the pyrimidines are somewhat simpler and consist of a single ringed structure, whereas the purines form two fused rings. Purines and pyrimidines differ according to the chemical groups that are attached to their rings. The genetic material of all living things is comprised of DNA. In viruses (which are not considered to be "alive"), the genetic material may be DNA or RNA. Except in the case of certain DNA viruses, DNA molecules consist of two polynucleotide chains twisted around each other to form a **double-helix**. In a DNA double helix, the nitrogenous bases of each chain project toward the center of the double helix. A purine of one strand is always associated with a pyrimidine of the other strand, the two strands held together in part by the formation of hydrogen bonds between the bases (Fig. 2-9).

Association of purines and pyrimidines in the center of the double helix is not random. Rather, the purine **adenine** (A) on one chain always faces the pyrimidine **thymine** (T) on the other chain. Similarly, **guanine** (G) is always matched with **cytosine** (C). Implicit in this organization is the fact that the sequence of nitrogenous bases of one chain of DNA determines the base sequence of the other chain. For example, if the sequence is A-G-C-A-T-C-T-G-A on one chain of the double helix, then on the other chain the base sequence is T-C-G-T-A-G-A-C-T. Thus, by knowing the base sequence of only one of the two polynucleotide chains of a DNA double helix, the base sequence of the matching chain can readily be determined. Moreover, if the two strands of DNA were separated, each strand's base sequence could be used as a guide to creating a matching chain and thereby two new (and identical) double helixes. Therein lies the capacity of DNA for replication (Fig. 2-9).

RNA serves as the genetic material of certain viruses, but in animals and plants RNA plays an intermediary role in converting the genetic message of DNA into a cell's proteins. Unlike DNA, RNA molecules consist of one polynucleotide chain; moreover, the pyrimidine, **uracil** occurs in place of thymine.

In addition to being the building blocks of nucleic acids, nucleotides are important in other ways. For example, **adenosine triphosphate (ATP)**, a compound of major importance in energy transfer during the body's chemical reactions (see Chapter 4), consists of a purine (i.e.,

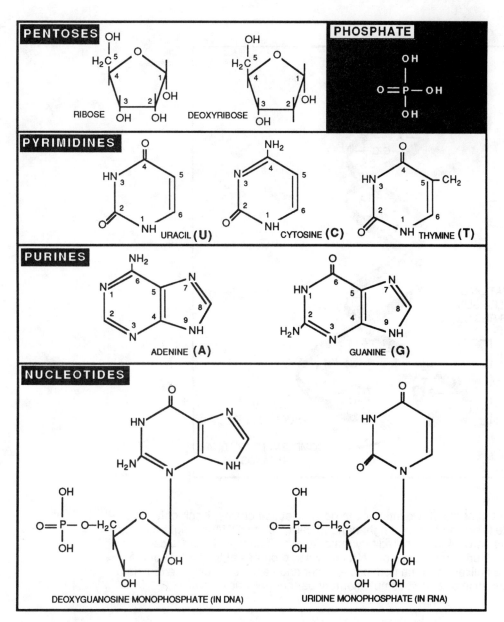

Figure 2-7
Nucleotides. Each nucleotide is comprised of a pentose sugar, a purine or pyrimidine, and a phosphate group. Only two of the eight different nucleotides that regularly occur in DNA and RNA are shown in the lower portion of the figure.

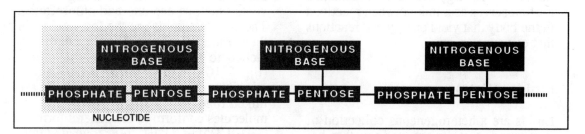

Figure 2-8
The organization of a polynucleotide chain. Usually, RNA consists of a single chain, whereas DNA consists of two intertwined chains (i.e., a *double helix).*

19

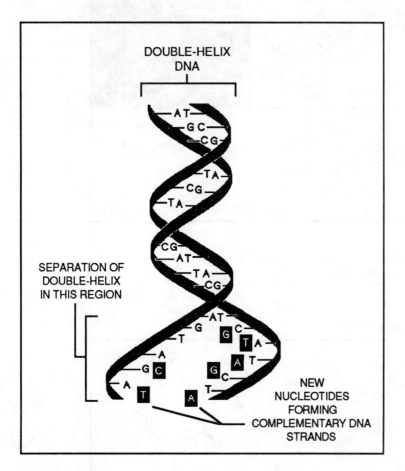

Figure 2-9

DNA. DNA consists of two interwoven polynucleotide chains. Each poly-nucleotide takes the form of a right-handed helix, so that DNA is a right-handed **double helix**. DNA is the hereditary material of all human cells and is capable of replication. During replication of DNA (shown in progress at the bottom of the figure), the helixes separate from one another and each helix acts as a template for the formation of a new, complementary partner polynucleotide.

adenine), ribose, and three phosphate groups. When we say that ATP is important in energy transfer, we mean that it often serves as a link between reactions in the body that yield energy and reactions that consume energy.

Lipids (Fats)

Lipids are a heterogeneous collection of compounds that are grouped together be-cause they share a common physical prop-erty; namely, they are insoluble in water but are soluble in a variety of organic sol-vents (such as acetone and chloroform). The simplest lipids are the **fatty acids**; in these molecules, a carboxyl group is at-tached to a long **hydrocarbon chain** (Fig. 2-10). In the *saturated* fatty acids, all of the carbon atoms of the chain are linked together by *single* bonds, and the molecules conform to the general formula $CH_3-(CH_2)n-COOH$. In *unsaturated* fatty

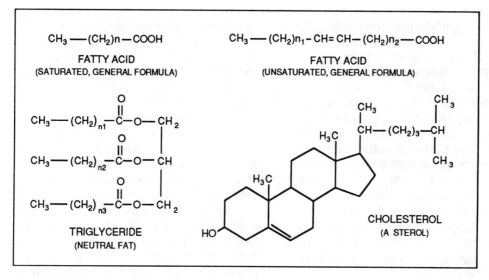

Figure 2-10
Lipids. Shown here are the general formulas for saturated and unsaturated fatty acids and triglycerides. Cholesterol is one of many different sterols found in the body.

acids, two or more of the carbon atoms of the hydrocarbon chain are linked by *double* bonds. Also common in cells and tissues are the **triglycerides** (also known as **neutral fats**) which consist of 3 fatty acids joined to a molecule of **glycerol** (Fig. 2-10). A variety of other compounds are also classified as lipids including **sterols** such as **cholesterol** (Fig. 2-10), certain **steroid hormones** and water-insoluble **vitamins** (e.g., *vitamin D*, found in milk).

METABOLISM AND ENZYMES

During a **chemical reaction**, one or more compounds (called **reactants**) are converted into one or more *different* compounds (called **products**). The changes in the reactants generally take the form of *rearrangements* of atoms, *addition* of atoms, or *removal* of atoms. In the living body, the passage of but a single second of time is accompanied by thousands of different chemical reactions, which collectively are called the body's **metabolism**.

Most of the chemical reactions that characterize the body's metabolism would

not proceed at all (or at best would proceed very slowly) were it not for the presence in the body of special **catalysts** that cause the chemical reactions to occur much more rapidly. All but a very small number of these catalysts are proteins called **enzymes**. Thus, it is the body's enzymes that allow us through *anabolic* reaction pathways to transform nutrients from our environment into body substance or through *catabolic* reaction pathways to "fuel" the body's functional machinery.

Most enzymes are proteins and are usually very much larger than the reactants that they change. With specific exceptions such as the enzymes of the digestive system, most of the enzymes function intracellularly and are immobilized in cells by their attachment to cellular membranes. When reactants (or **substrate**) come in contact with an enzyme, the substrate becomes bound to the enzyme, thereby forming a transient complex called the **enzyme-substrate complex**. The substrate is then converted to product which is quickly released from the enzyme. The enzymes itself is not changed at all during the reaction that it catalyzes, and at the

21

reaction's completion the enzyme is again available to catalyze another reaction. It is not unusual for an enzyme to be able to catalyze many thousands of successive (but identical) reactions in one second.

Substrate molecules attach to a particular portion of the enzyme called the **active site**, and it is the shape and chemical nature of the active site that determines the properties of the enzyme. The active site may not only bind the substrate but may also bind a **cofactor** (e.g., a metal ion such as Ca^{++} or Mg^{++}; see below) that assists the enzyme in its catalysis. The active site of an enzyme usually occupies only a small portion of the enzyme's surface.

Enzymes range in molecular weight from about 5,000 to many millions of daltons. The biological activity of an enzyme (i.e., its ability to perform its catalysis) is usually lost at temperatures much above body temperature. This loss of activity is known as **denaturation**. The heat sensitivity of an enzyme is related to the molecule's large size and the relatively weak forces (e.g., hydrogen bonds) that maintain its complex and intricate shape. Enzymes are also denatured by small changes in the pH of their surroundings. This is why the digestive enzyme *pepsin* functions properly only in the acidic environment of the stomach, whereas the *salivary enzymes* function properly only in the alkaline saliva of the mouth.

As noted above, the catalytic actions of enzymes often require the participation of smaller molecules called cofactors. There are three kinds of cofactors: (1) **coenzymes**, which are organic compounds and are transient in their association with the protein part of the enzyme; (2)

prosthetic groups, which also are organic compounds but which are permanently bound to the protein, and (3) **metal ions**. Coenzymes and cofactors are usually comprised (at least in part) of **vitamins** (hence the importance of vitamins to the body's metabolism).

The catalysis displayed by most enzymes is highly "specific." For example, an enzyme that catalyzes a chemical change in glucose will not bring about a similar change in some other sugar. The specificity of enzymes is due to the rather constrained nature of the interaction between the enzyme's active site (which has a particular shape and displays specific physical and chemical properties) and the substrate of that enzyme (which also has a particular shape and displays specific properties). Thus, the geometry of the active site determines what substrate molecules will be bound. Many enzymes catalyze reactions in *both* directions. That is, an enzyme that converts compound "A" to compound "B" may also be able to convert compound "B" to compound "A." The direction in which the catalysis takes place is determined by the relative amounts of "A" and "B" that are present (e.g., when there is an excess of "A," the enzyme converts "A" to "B," and vice-versa).

Most of the body's metabolism is intracellular, and therefore most of the enzymes of the body are localized within (or on the surfaces of) different cellular components. Consequently, let us now turn to a review of the organization of cells and their functional components.

CELLS

The body is comprised of billions and billions of structural and functional units called **cells**. Despite the small size of cells, much has been learned about cell structure and organization using various forms of *light* and *electron* microscopy. Although all cells appear to have certain features in common, all cells are not exactly alike. Rather, cells occur in a variety of shapes and sizes and exhibit a broad range of physiological properties. This range of cellular attributes is directly related to the variety of tissues in which cells are found and to the diversity of functions that the various cells and tissues of the body must perform. For example, not only do human liver cells and nerve cells look quite different, they also exhibit many differences in their physiology. Notwithstanding the features that distinguish one cell type from another, there are also features that all human cells share. Let's begin this discussion of cells by considering the general features of cell structure (Fig. 3-1).

THE CELLULAR ORGANELLES

The internal structures of a cell play specific roles in the cell's overall function and behavior. This is not to say that each part of a cell behaves independently of the other parts; rather, the activity of each cell structure is influenced by (and at the same time has an influence on) other cell parts. The discrete functioning parts that comprise a cell are also referred to as the cell's **organelles**.

Plasma Membrane

At its surface, a cell is bordered by a membrane called the **plasma membrane** (or **plasmalemma**) composed for the most part of thousands of protein and lipid molecules (Fig. 3-1). Most substances that enter or leave the cell must pass through this membrane. Since (1) most human cells obtain their nutrients from and dispatch their wastes into the external cell surroundings, and (2) the greater the ratio of surface area of the plasma membrane to cell volume, the greater can be the rate of exchange between a cell and its surroundings, a cell's shape is often modified in order to maximize the *surface area:volume ratio*. In the case of some cells (e.g., red blood cells), this is achieved by changing the overall shape of the cell to that of a flattened disk. However, in many cells (e.g., absorptive cells of the digestive tract), the surface area:volume ratio is increased through the formation of numerous tiny projections of the plasma membrane called **microvilli**.

The plasma membrane actively regulates the passage of materials between the cell and its surroundings. In some tissues, the plasma membrane is involved in intercellular communication (e.g., nerve tissue), in which materials and information

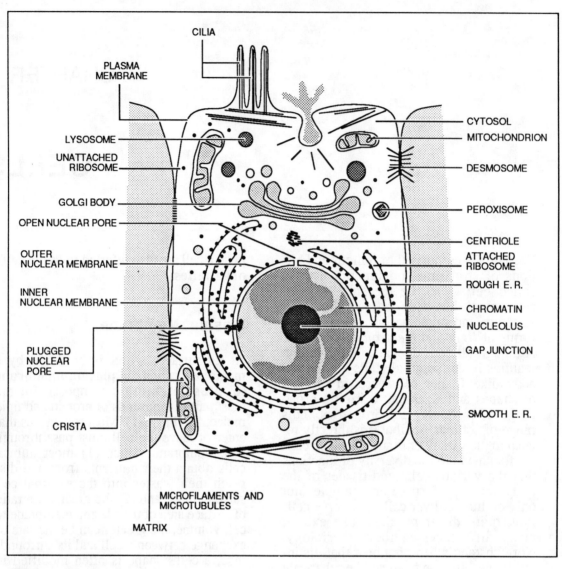

CILIA

PLASMA
MEMBRANE

CYTOSOL

MITOCHONDRION

LYSOSOME

DESMOSOME

UNATTACHED
RIBOSOME

GOLGI BODY

PEROXISOME

OPEN NUCLEAR PORE

CENTRIOLE

OUTER
NUCLEAR MEMBRANE

ATTACHED
RIBOSOME

ROUGH E. R.

INNER
NUCLEAR MEMBRANE

CHROMATIN

NUCLEOLUS

GAP JUNCTION

PLUGGED
NUCLEAR
PORE

CRISTA

SMOOTH E. R.

MICROFILAMENTS AND
MICROTUBULES

MATRIX

Figure 3-1
A generalized human cell.

are sent from one cell to another. In most of the body's tissues, it is not unusual to observe special junctions between the plasma membranes of neighboring cells. These take the form of **tight junctions**, **desmosomes**, and **gap junctions**.

Chemically, the plasma membrane consists of protein, lipid, and carbohydrate molecules. The lipid is primarily phospholipid and according to the **fluid mosaic model** of membrane structure is organized into two layers (Fig. 3-2). Protein

molecules are associated with the membrane in either of two ways. They may penetrate one or both lipid layers or they may be loosely attached to the outer or inner lipid surfaces. Carbohydrate is associated only with the outer half of the membrane, where it is chemically bonded to membrane protein and/or lipid. The rather simple appearance of the plasma membrane when examined by microscopy belies the heterogeneity of its chemical organization. The kinds of proteins, lipids, and carbohydrates that make up the membrane vary in different regions of the membrane. This heterogeneity may be illustrated using the plasma membrane of a liver cell as an example (Fig. 3-3). Some areas of a liver cell's plasma membrane face the plasma membranes of neighboring liver cells; other areas face the bile channels into which bile and other substances produced in the liver cell are secreted. Still other portions of the plasma membrane face the surfaces of capillaries from which substances are absorbed. Each of these regions of the plasma membrane is differently composed and differently organized and, in fact, is continually undergoing change and reorganization.

The Cytoskeleton, Microtrabecular Lattice, and Endoplasmic Reticulum

Radiating through many cells of the body are the filaments and microtubules of the **cytoskeleton** and **microtrabecular lattice**. These structures give shape and form to the cell and are also involved in cell movement. Also radiating through the cytoplasm of most cells is an extensive network of branching and fusing membrane-limited channels (or *cisternae*) col-

lectively called the **endoplasmic reticulum** (usually abbreviated **ER**). The ER membranes (as are all other cellular membranes) are comprised of protein and lipid molecules that are organized in much the same manner as in the plasma membrane.

The membranes of the endoplasmic reticulum divide the cell into two phases: the **intracisternal** phase and the **cytosol**. The intracisternal phase consists of the material enclosed within the cisternae of the endoplasmic reticulum, while the cytosol surrounds the ER membranes. In the cytosol are large numbers of small particles called **ribosomes**. These particles, which are comprised of RNA and protein molecules, are distributed along the cytosolic surface of the endoplasmic reticulum ("attached" ribosomes) and are also free in the cytosol ("free" ribosomes). There is some evidence that the free ribosomes are interconnected by fine filaments of the microtrabecular lattice. Ribosomes synthesize the cell's proteins. Endoplasmic reticulum with attached ribosomes is called *rough* ER (RER), the membranes of the RER typically being sheet-like. Endoplasmic reticulum without attached ribosomes is called *smooth* ER (SER), the membranes usually forming a network of branching and fusing tubes. Smooth ER appears to be the site for lipid synthesis.

Figure 3-2
The **fluid-mosaic** model of membrane structure.

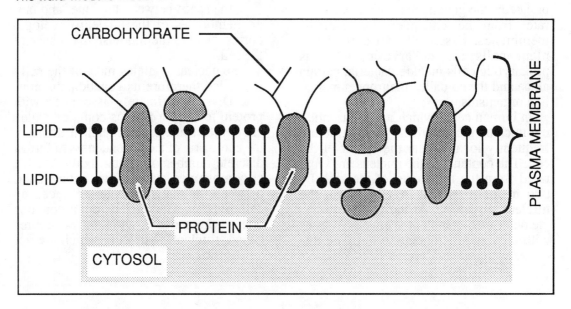

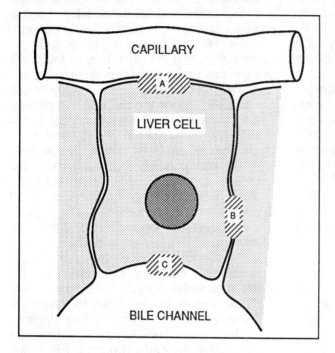

Figure 3-3
The various faces (i.e., "domains") of a liver cell's plasma membrane. *Domain A* faces the bloodstream, *domain B* faces neighboring liver cells, and *domain C* faces the bile channels into which bile is secreted. In each of these domains, the membrane's protein and lipid composition is different.

Nucleus

All cells (human or otherwise) may be assigned to one of two major categories: **eukaryotic** (i.e., "true nucleus") and **prokaryotic** (i.e., "before nucleus"). In eukaryotic cells, the genetic material is separated from the rest of the cell by a membranous envelope that forms a structure called the cell **nucleus**. In prokaryotic cells, the genetic material is *not* separated from the remainder of the cell by membranes. Essentially all animal and plant cells are eukaryotic, whereas prokaryotic cells include bacteria, certain algae and the so-called pleuropneumonia-like organisms.

In human cells (which are eukaryotic), the nucleus is a relatively large structure frequently located near the center of the cell. The contents of the nucleus are separated from the remainder of the cell by two membranes that together form the **nuclear envelope**. At various positions, the outer membrane of the envelope fuses with the inner membrane to form the **nu-**clear pores. These pores provide a direct path between the cytosol and the contents of the nucleus. Some of the nuclear pores may be plugged by a granular material. The outer nuclear membrane may have ribosomes attached to its cytosol side and may also merge with the membranes of the endoplasmic reticulum. Since the ER is continuous at certain points with the plasma membrane, the **perinuclear space** (i.e., the space between the inner and outer membranes of the nuclear envelope) corresponds to the intracisternal phase of the cell.

The nucleus contains most of the cell's DNA and therefore its genetic apparatus. The DNA is intimately associated with protein to form a fibrous complex called **chromatin**. Just prior to nuclear division, the chromatin of a cell condenses to form a discrete number of visible bodies called **chromosomes,** which are apportioned between the daughter cells produced by cell division. The nucleus often contains one or more dense, granular structures called **nucleoli**. Nucleoli, which are not bounded by a membrane, contain concentrations of particles that ultimately are incorporated into the cell's ribosomes.

Mitochondria

Distributed through the cytosol are large numbers of organelles called **mitochondria**. The number of mitochondria per cell is quite variable; for example, sperm cells have fewer than 100 mitochondria, kidney cells generally contain less than 1000, and liver cells may contain several thousand. Mitochondria are oval bodies enclosed by two distinct membranes called the *outer* and *inner* membranes. The inner membrane separates the organelle's volume into two phases: the **matrix**, which is a gel-like fluid enclosed by the inner membrane, and the fluid-filled **intermembrane space** between the inner and outer membranes. The inner membrane has a much greater surface area than the outer membrane because it possesses folds that extend into the matrix. These projections are called **cristae**.

Mitochondria play several different roles in a cell. Probably the most important is the mitochondrion's capacity to break down certain organic acids into carbon dioxide and water, and in so doing produce chemical energy for the cell in the form of ATP.

Golgi Bodies

Golgi bodies are organelles that consist of sets of smooth, flattened cisternae stacked together in parallel rows and surrounded by vesicles of various sizes. These organelles play a variety of functions including (1) the packaging of chemical substances that are to be secreted from the cell, (2) the chemical modification of proteins that have been synthesized by the cell's ribosomes, (3) the synthesis of certain of the cell's polysaccharides, and (4) the production of membrane components for the cell's plasma membrane and other membranous organelles.

Lysosomes and Microbodies

Most human cells contain small vesicular structures called **lysosomes**. Lysosomes are bounded at their surface by a single membrane and contain quantities of various hydrolytic enzymes capable of digesting proteins, nucleic acids, polysaccharides, and other materials. Under normal conditions, the activity of these enzymes is confined to the interior of the organelles and is therefore isolated from the surrounding cytosol. However, if the lysosomal membrane is ruptured, the released enzymes can quickly degrade the cell. Among their various roles, lysosomes take part in the intracellular digestion of particles that are ingested by the cell during *endocytosis* and the intracellular scavenging of worn and poorly functioning organelles.

Lysosomes are related to another family of organelles called **microbodies**. Among human cell microbodies, the most common are the **peroxisomes**. These small organelles, contain a number of enzymes whose functions are related to the breakdown of potentially harmful peroxides that are produced during a cell's metabolism.

Cilia

The surfaces of many human cells possess rows of hairlike extensions called **cilia**. These organelles serve to move a substrate across the cell surface (such as the movement of mucus in the respiratory tract, or the movement of an egg cell during its passage through the oviduct from the ovary to the uterus). Each cilium is covered by an extension of the plasma membrane. Internally, cilia contain an array of microtubules that run from the base of the organelle (just below the plasma membrane) toward the cilium's tip.

BACTERIA

Bacteria are small, prokaryotic microorganisms typically about the size of a

27

mitochondrion of a human cell and are the causes of a number of human diseases and infections. Bacterial cells are generally enclosed within a wall formed from protein and lipid as well as poly-saccharide. The wall lies external to the cell's plasma membrane and its content of a particular protein-carbohydrate complex is the basis of the classification of bacteria as "gram-positive" or "gram-negative." Plasma membrane infoldings in gram-positive bacteria give rise to structures called **mesosomes**, which play a role in the division of the cell. In some bacteria, there is a layered arrangement of membranes within the cytoplasm. However, there are no structures comparable to the endoplasmic reticulum. Bacteria contain large numbers of ribosomes, but most of these organelles are free in the bacterial cytosol; some ribosomes may be attached to the interior surface of the plasma membrane.

In bacteria, the nuclear material is not separated from the cytosol by membranes, as it is in human (and other eukaryotic) cells. However, the nuclear material of a bacterial cell is usually concentrated in a specific region of the cell referred to as a **nucleoid**. Most of the hereditary material of the cell is carried by a single, circular chromosome; however, a small amount of genetic material is also present in small circular bodies called **plasmids**.

PLEUROPNEUMONIA-LIKE ORGANISMS

Pleuropneumonia-like organisms (also called mycoplasmas) are prokaryotic and are the smallest and simplest of all cells. They are the causes of a number of diseases in humans and other animals. Pleuropneumonia-like organisms are bounded at their surfaces by a membrane composed of proteins and lipid, but there is no cell wall. Internally the cell's composition is more or less diffuse. The only microscopically discernible features within the cell are its genetic complement (which consists of a double-helical strand of circular DNA) and a number of ribo-somes.

VIRUSES

Viruses are not cells, and although they are smaller even than most prokaryotic cells, they are diverse in size and in organization. Viruses are the sources of many infections and diseases; even the smallest of cells (e.g., bacteria and mycoplasmas) are subject to infection by viruses. Among the viruses that attack human cells, the most notorious are the those that cause smallpox, chicken pox, rabies, poliomyelitis, mumps, measles, influenza, hepatitis, AIDS (acquired immune deficiency syndrome), and the "common cold." Even certain leukemias and cancers are of viral origin.

Most viruses are either rod-shaped or quasi-spherical and contain a nucleic acid **core** surrounded by a specific geometric array of protein molecules that form a coat or **capsid** (Fig. 3-4). In many viruses, a lipoprotein envelope surrounds the capsid (e.g., influenza virus, herpes virus, and smallpox virus).

The Cycle of Infection of A Virus

Solitary viruses do not carry out metabolism and are incapable of reproducing. Proliferation of viruses requires a **host cell**, and in its simplest and most direct form takes the pattern illustrated in the stages of Figure 3-5. One or more viruses attach to specific sites on the plasma membrane of the host cell. Following attachment, the viral nucleic acid is inserted through the plasma membrane into the host's cytosol. Once inside the host cell, the viral nucleic acid redirects the metabolism of the host so that new viral proteins and new viral nucleic acids are formed. These viral components combine in the host to form large numbers of new viruses that exit the cell by disruption of the cell's plasma membrane. Some viruses

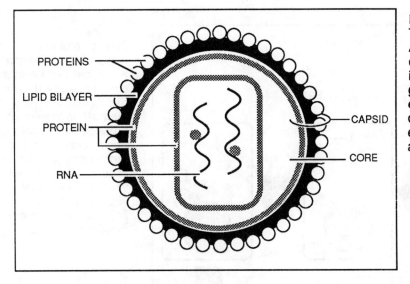

Figure 3-4
The **AIDS** virus.
Although the organization of viruses is quite variable, in most viruses the viral **genome** (consisting either of DNA or RNA, depending on the virus) is enclosed within a protein and lipid-rich **capsid.**

exit the host by budding off, enclosed in a small piece of the host cell's plasma membrane; this does not lyse the host cell. The cycle of infection then repeats itself.

On some occasions and only for certain viruses, the injected nucleic acid does not cause proliferation and release of new viruses. Instead, the injected nucleic acid is incorporated into the host's genetic material, and the host cell continues to function in its normal manner. However, duplication of the host's genetic material prior to cell division is accompanied by duplication of the incorporated viral nucleic acid. Several generations of cells may be produced, each containing a copy of the viral nucleic acid. The dormant viral nucleic acid within the host is referred to as a **provirus.** Sooner or later, in one of the generations of host cells, the provirus nucleic acid will begin to direct the replication of new viruses, and this in turn will lead to the release of new, infective virus particles.

Viral Nucleic Acids

Viruses may be assigned to one of two major groups: **DNA viruses** (i.e., viruses in which the hereditary material is DNA) and **RNA viruses** (viruses in which the hereditary material is RNA).

The DNA viruses include those that cause chicken pox, herpes blisters, infectious mononucleosis, and shingles. Among the RNA viruses are those that cause polio, mumps, measles, influenza, AIDS, and colds. Some RNA viruses are also believed to be the causes of certain cancers.

VIROIDS

Viroids are also agents of disease but they are much smaller than viruses, consisting of no more than a single strand of RNA. The RNA is not enclosed in any structure, and except during infection of a host cell, is not associated with any other chemical substances. At the present time, viroids are only suspected of being the agents of certain diseases in animals, whereas they are known to produce a number of plant diseases.

PRIONS

Generally, the sources of human infection are either viroids, viruses, or bacteria. What these infectious agents have in common is that their identity is defined by the nucleic acid that each carries. It now appears that there may be an exception to

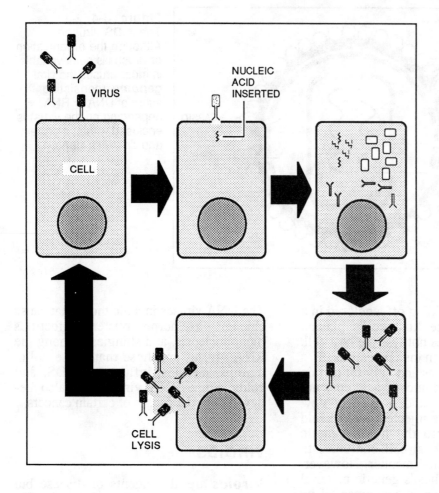

Figure 3-5
The cycle of infection of a virus. Viruses attach to a host cell and insert their core nucleic acid. The metabolic machinery of the host is then used to produce large numbers of new viral cores and protein coats, which are assembled to form whole viruses. Eventually, the viruses break out of the host and repeat the cycle of infection.

this rule. *Scrapie* (a disease of goats and sheep) and a disease of the nervous system in humans (called *Creutzfeldt-Jakob* disease) appear to be caused by agents consisting only of protein. The presumed agents of these diseases are called **prions**.

FROM CELLS TO ORGAN-SYSTEMS

Cells/Tissues

In the human body, groups of similar cells performing the same function are grouped together to form **tissues**. There are four major types of tissues in the body; these are (1) **epithelial** tissues, (2) **muscle** tissues, (3) **nerve** tissues, and (4) **connective** tissues.

Epithelial tissues line the body surfaces, such as the skin and the walls of the digestive and respiratory tracts. Their roles are principally protective, but much absorption of chemicals into the body or the secretion of materials from the body involve the actions of epithelium. The principal function of muscle tissue is that of contraction, and this tissue is responsible for all body movements (internal and external). Nerve tissue comprises such structures as the brain and spinal cord; this tissue is specialized for the conduction and transmission of signals (e.g., nerve impulses) from one part of the body to another. Connective tissue serves diverse functions. Some forms of connective tissue join other tissues together. For example, tendons connect muscle tissue to bone. Other forms, such as cartilage, give support to the body. Fat (or adipose) tissue serves as storage sites.

Tissues/Organs

A group of different tissues held together and working as a unit in order to provide a basic body function is called an **organ**. For example, the heart is an organ comprised of muscle tissue, connective tissue, nerve tissue, and epithelial tissue. All of the tissues of this organ contribute in some manner to the organ's principal function, namely to pump blood through the body.

Organs/Organ-Systems

Groups of organs that serve an overall function–even though the organs may be located in diverse regions of the body–comprise an **organ-system**. For example, the heart is just one organ of the organ-system known as the circulatory system; the other organs of the circulatory system include arteries and veins. In a like manner, the brain is just one organ of the organ-system known as the nervous system; the stomach is just one organ of the organ-system known as the digestive system, and so on.

In succeeding chapters of this book, we will consider the physiology of the body by surveying the organization and functions of each of the body's organ systems.

~~~~~~~~~~~~~~~~~~~~~~~~~~~~~~~

THE PHYSIOLOGY OF MUSCLE

The body's muscles contain a specialized form of tissue whose cells have the capacity to contract, that is, to physically shorten. It is the muscle cells' capacity for **contraction** that is responsible for essentially all movements associated with the body, whether the movements are overt and external (e.g., movements of the arms and legs) or internal (e.g., movements of the digestive organs and the beating of the heart).

TYPES OF MUSCLE TISSUE

There are *three* different types of muscle tissue found in the body; these are called (1) **striated** muscle tissue, (2) **smooth** muscle tissue, and (3) **cardiac** muscle tissue. Among the three, much more is known about the chemistry, organization, and physiology of striated tissue than is known about the others, primarily because striated tissue is so much more accessible to study. Consequently, much of this discussion of muscle will center around the organization and physiology of striated tissue. Let's begin our consideration of muscle by identifying the principal roles and dispositions of these tissues in the body.

The word "striated" means "striped." Striated muscle is given this name because when the tissue is examined with either a light or electron microscope, the cells that comprise the tissue appear to be have stripes (striations) running across their widths (i.e., the stripes are oriented perpendicular to the long axes of the cells). Because most striated muscle is attached to bone and moves the skeleton, striated tissue is sometimes referred to as "skeletal" muscle. However, this term is misleading because not all striated muscle is associated with the skeleton. For example, the tongue and the diaphragm (a major muscle of respiration) are not attached to bone.

Smooth muscle tissue is called "smooth" because it lacks the stripes characteristic of striated cells. Smooth muscle tissue is associated with the body's internal organs. For example, it is abundant in the walls of the digestive organs (e.g., the stomach and intestine), the urinary bladder, and the walls of the major arteries and veins. Like striated muscle, smooth muscle is associated with movement but the movement is internal and more subtle. Much of this movement goes on without conscious awareness.

As its name implies, "cardiac" muscle is found in the heart. Indeed, the heart is the only organ that contains this type of muscle tissue. The major role of cardiac muscle is to provide the continuous and rhythmic contractions that propel the blood through the circulatory system.

33

ORGANIZATION OF STRIATED MUSCLE

The organization of a typical striated muscle can be illustrated using the **biceps** muscle as an example (Fig. 4-1). The biceps muscle is a muscle of the upper arm whose principal function is to draw the lower arm (i.e., the "forearm") upward, toward the shoulder. The action of this muscle demonstrates a characteristic that is typical of the body's striated muscles, namely that during contraction one end of the muscle undergoes very little change in position, whereas the other end moves a considerable distance. In the case of the biceps muscle, the end that undergoes little movement (called the muscle's **origin**) is the end attached to the shoulder region, whereas the end of the biceps that attaches to the forearm (called the muscle's **insertion**) moves several inches. The origins and insertions of the muscles that move the arms, fingers, legs, toes, head, jaw, and eyes are readily identified. However, for some muscles (such as those of the back), the origin and insertion are not so obvious.

Figure 4-1
The **biceps** muscle.

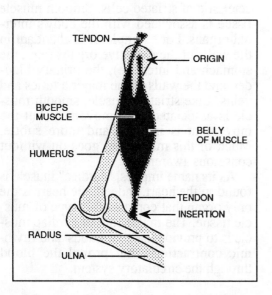

The movements of the biceps also illustrate another important aspect of muscle action; that is, *muscles pull, they do not push*. The action of a muscle can be likened to the that of a tow-rope, in that the rope can be used to pull an object along, but it can't be used to push an object along. Even familiar pushing actions such as pushing yourself away from the dinner table or pressing down on the car's accelerator peddle are the results of muscle shortening (i.e., contracting). In the case of pushing yourself away from a table, the contraction of another muscle of the upper arm called the **triceps** rotates the forearm *away* from the shoulder (the elbow acting as the pivot), thereby extending the arm. This action is opposite to that of the biceps. Indeed, there are many examples in the body of pairs of *antagonistic muscles*.

The gross actions of a muscle are founded on the individual actions of the thousands (millions in the larger muscles) of contractile cells that make up the muscle. In striated muscles, the muscle cells are long, threadlike structures oriented along the axis that leads from the origin to the insertion. Because they are threadlike, striated muscle cells are also called striated muscle **fibers**; that is, the term "fiber" is equated with "cell." In the longer muscles (e.g., the muscles of the arms and legs), individual fibers may be several inches long and are among the largest cells in the body. The muscle fibers that comprise a striated muscle do not extend the entire length of the muscle. Instead, the ends of a muscle are formed by a **connective tissue** called **tendon** that serves to connect the contractile portion of the muscle to the bone on which it pulls.

The internal organization of a striated muscle is shown in Figure 4-2, which depicts a cross-section through the muscle's "belly." The muscle is enclosed by a band of connective tissue called **epimyseum**. Internally, the individual fibers are arranged in large bundles called **fasciculi** (singular = **fasciculus**). Each fasciculus is covered by a layer of connective tissue called **perimyseum**. Larger muscles typically have a number of fasciculi. Blood vessels that nourish the muscle tissue and nerves that control the muscle's actions run between the fasciculi. Within each fasciculus, groups of muscle fibers are embedded in yet another form of connective tissue called **endomyseum**. The fibers seen in Figure 4-2 are depicted in cross-section; to appreciate their remarkably complex internal organization, we need to withdraw one of these fibers from the muscle, and turn it sideways so that we can view its length. This arrangement is shown in Figure 4-3.

Figure 4-2
Internal organization of a muscle.

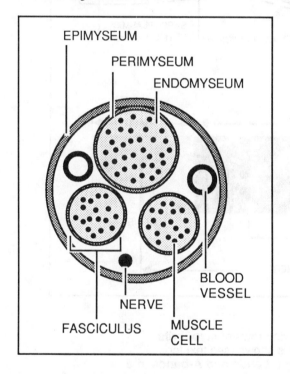

Because most striated fibers are exceedingly long, Figure 4-3 depicts only a small segment of the whole cell. The unusual length of these cells stems from the fact that during embryonic development, the muscle fiber is formed by the end-to-end fusion of many smaller cells, thereby forming a continuous tubelike structure. This also explains why striated muscle fibers are *multinucleate* (i.e., they have

many nuclei). The plasma membrane that encloses each cell is called the **sarcolemma**; in addition to its exceptional length, the sarcolemma is characterized by numerous porelike invaginations that extend into the interior of the cell (i.e., into the cytoplasm or **sarcoplasm**) at right angles to the cell's long axis. Within the cell, these invaginations form a highly branched network called the **T-system** (i.e, "T" = "transverse") or **sarcoplasmic reticulum**. The interior of the cell is also characterized by large numbers of rodlike structures called **myofibrils**. The myofibrils are the cell's contractile units; that is, they are the structures that directly bring about the cell's physical shortening. As seen in Figure 4-3, each myofibril reveals areas of different density. The alternating light and dark areas (the stripes or striations referred to earlier) are respectively called the **I** (i.e., **isotropic** [which means transparent]) and **A** (i.e., **anisotropic** [which means opaque]) **bands**. In the lower half of Figure 4-3, the alternating A and I bands are depicted in greater detail. At the center of each I-band, there is a dark line called a **Z-line**. Within each A-band, there is a region that appears somewhat less dense than the remainder of the band; this is called an **H-zone**. Finally, at the center of an H-zone is the darker **M-line**. Those segments of each myofibril that extend from one Z-line to the next are termed **sarcomeres**. Thus, a myofibril may be thought of as a succession of sarcomeres.

Arrangement Of Filaments Within A Fibril

The myofibrils of a muscle fiber are comprised of even finer, threadlike structures called **myofilaments** (Fig. 4-4). There are two kinds of myofilaments in a myofibril: (1) **thin filaments** composed of the protein **actin**, and (2) **thick filaments** composed of the protein **myosin**. One end of each thin myofilament is anchored in a Z-line, while the other end projects toward the center of the sarco-

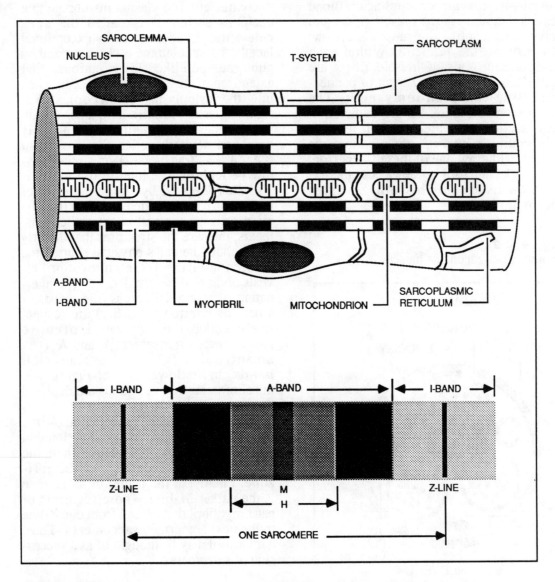

Figure 4-3
Top: Organization of a striated muscle fiber showing rows of **myofibrils**. Only a small portion of the total length of the fiber is shown (see text for explanation). *Bottom:* The pattern of alternating **I- bands** and **A-bands** in a single myofibril.

mere; the thin myofilaments are *isotropic* (light readily passes through them). The thick myofilaments are anisotropic (i.e., they absorb light) and are sandwiched between the thin filaments, overlapping their ends, but are not attached to the Z-lines. As seen at the bottom of Figure 4-4, which shows the filaments in cross-section, *six* thin filaments are equally spaced about each thick filament in an hexagonal array. Several hundred thick and thin myofilaments are bundled together to form

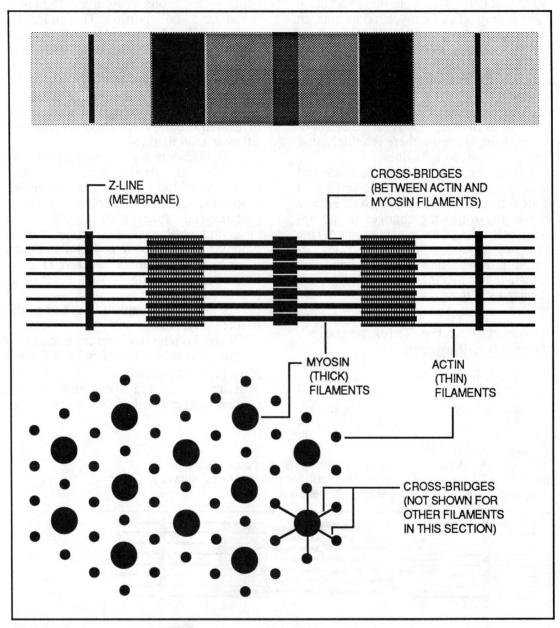

Figure 4-4
One end of each thin (**actin**) filament is anchored in a Z-line, while the other end projects toward the center of the sarcomere. Thick (**myosin**) filaments are sandwiched between thin filaments, but are not attached to the Z-lines. Cross-bridges connect overlapping portions of the filaments. Seen in cross-section (bottom), *six* thin filaments are equally spaced about each thick filament.

each myofibril. It is the light-absorbing properties and peculiar orientations of the thick and thin myofilaments that serve as the basis for the alternating dark (i.e., A) and light (i.e., I) banding pattern that characterizes each myofibril. Within a sarcomere, an A band extends from one end of a stack of thick filaments to the other end (Fig. 4-4). Note that in addition to including all of the myosin filaments, an A band includes a small portion of each of the thin filaments (i.e., the segments that overlap the thick filaments). That portion of each A-band that is devoid of actin filaments appears somewhat less dense than the remainder of the A-band and creates an H-zone. Finally, at the center of the myosin filaments, there is a thickening that gives rise to the M-line.

During muscle activity, the thick and thin myofilaments of the myofibrils of a contracting fiber slide past each other; as a result, the following changes in the appearances of each sarcomere occur: (1) the H-zones disappear, (2) the I-bands become narrower, and (3) the Z-lines are drawn closer together. These changes are depicted in Figure 4-5.

Chemistry of the Thick and Thin Myofilaments

The thick myosin filaments are formed from bundles of individual myosin molecules (Fig. 4-6). Each myosin mol-ecule is a fibrous protein containing a "head" and "tail" portion. The molecules are arranged with their tails parallel to each other and their heads projecting away from the long axis of the filament at intervals. No heads are present in the center of the filament, and this is the region that coincides with the an H-zone. The heads of the myosin molecules form *cross-bridges* with adjacent actin filaments.

Thin filaments are composed primarily of actin, but also present are small quantities of the proteins **tropomyosin**, **troponin**, and **actinin**. Soluble actin is a globular (i.e., spheroidal) protein, but in the actin myofilaments two chains of these globular molecules are coiled about each other to form a fibrous structure (Fig. 4-6). As depicted in Figure 4-6, it is the oar-like action of these cross-bridges that brings about filament sliding and is the basis of contraction.

Of special importance in the cyclic contraction and relaxation of muscle are the roles played by calcium ions (i.e., Ca^{++}). In a non-contracting fiber, these ions are concentrated in small extensions of the sar-

Figure 4-5
Top: During contraction, thin and thick filaments slide over each other, pulling the Z-lines closer together. *Bottom:* Simplified scheme showing one myosin filament and four actin filaments per sarcomere.

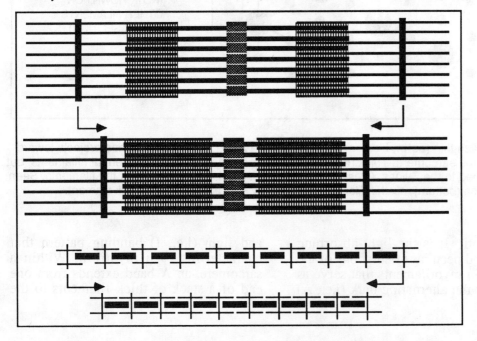

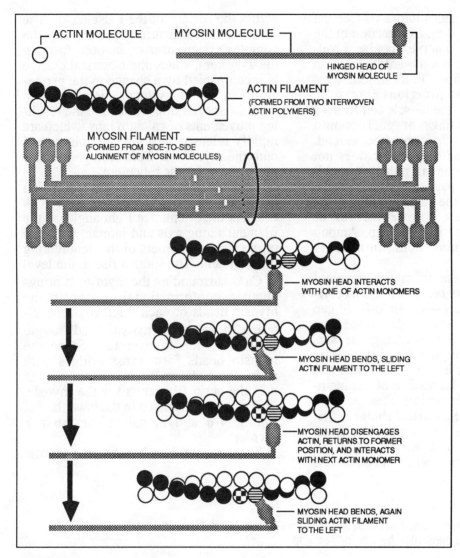

Figure 4-6
Thin filaments are formed from two interwoven chains of globular actin molecules. Thick filaments are formed from bundles of fibrous myosin molecules arranged with their "tails" parallel and their "heads" projecting away from the filament's long axis. Myosin heads form *cross-bridges* with adjacent actin filaments. The oar-like action of the cross-bridges brings about filament sliding (lower half of figure).

coplasmic reticulum. It is the release of Ca⁺⁺ from the sarcoplasmic reticulum that initiates filament sliding (see below).

The Contraction/Relaxation Cycle

Several mechanical and electrical condi-tions characterize the status of a muscle cell prior to the onset of its contraction. The sarcolemma is **polarized**, its outside surface bearing *positive* charge and its in-side surface bearing *negative* charge. In-ternally, the cross-bridges between the myosin heads and the actin filaments of each myofibril are only weakly interacting, with each myosin head oriented at right angles to the filament's long axis (see the lower half of Fig. 4-6). Contraction of the cell is initiated by a nerve impulse travelling along a nerve cell toward its junction with the muscle fiber. In these junctions, called **myoneural junctions** (Fig. 4-7), the sarcolemma of the muscle cell forms a number of infoldings or pockets into which branches of the nerve fiber extend. The surface of the nerve fiber does not touch the sarcolemma; rather, a small fluid-filled gap separates the two. As the impulse reaches the nerve endings, the nerve cell secretes a **neurohumor** or **neurotransmitter** into the gap. Among the most common neurohumors is **acetylcholine**.

After its release from the terminal branches of the nerve fiber, the acetylcholine diffuses across the fluid-filled gap that separates the nerve cell's surface from the sarcolemma. On reaching the sarcolemma, the acetylcholine attaches to specific receptors in the sarcolemma and initiates the **depolarization** of the membrane. This depolarization takes the form of a reversal in the electrical charge distribution across the sarcolemma. Depolarization begins at the myoneural junction but quickly spreads from there in all directions

across the surface of the muscle cell. The wave of depolarization spreads into the complex system of invaginations forming the T-system, where the electrical change is accompanied by a change in the permeability of the T-system's membranes to ions. Of special importance are the resulting movements of calcium ions, which are rapidly released into the sarcoplasm and onto the myofibrils from storage vesicles of the sarcoplasmic reticulum.

As noted above, prior to the contraction of a fiber, the myosin heads of each thick filament are oriented at right angles to the filament's long axis and interact with one of the actin monomers of the neighboring thin filament. The sudden rise in the level of Ca^{++} surrounding the myofibrils brings about a conformational change in the myosin heads of each thick filament, altering to $45°$ each myosin head's angle relative to the myosin tail. Since the myosin heads form cross-bridges with actin, the conformational change acts to pull the actin filament over the myosin. This change is depicted in the lower half of Figure 4-6 and is called the **power stroke**.

The cross-bridges between the myosin

Figure 4-7
The **myoneural junction**. Note that the surfaces of the nerve fiber and the muscle cell are separated by a fluid-filled gap.

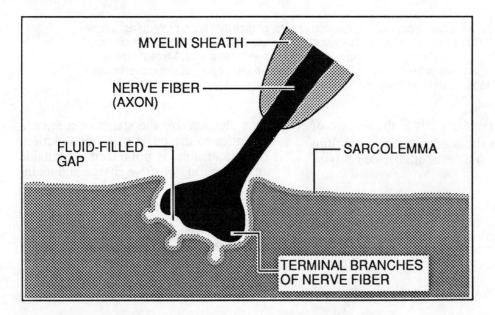

MYELIN SHEATH

NERVE FIBER (AXON)

FLUID-FILLED GAP

SARCOLEMMA

TERMINAL BRANCHES OF NERVE FIBER

heads and the actin filament break and the myosin heads return to their former 90º orientations. Now, however, each myosin head is positioned next to a "new" actin monomer (i.e., the next actin monomer in line). Again, the transition from 90º to 45º occurs, the actin filament pulled another tiny distance over the myosin. The action is repeated over and over until maximum shortening of the muscle fibril is attained, the actions occurring simultaneously in all of the myofibrils of the cell.

Meanwhile, back at the myoneural junction, the muscle fiber releases an enzyme into the junction that degrades the neurotransmitter. The enzyme that specifically acts on acetylcholine is called **acetylcholinesterase**. Destruction of all of the acetylcholine at the myoneural junction by acetylcholinesterase permits the sarcolemma at the junction to **repolarize** (i.e., to restore the former electrical charge distribution across the membrane). Repolarization then spreads like a wave over the sarcolemma from the myoneural junction and passes into the T-system. Repolarization of the sarcoplasmic reticulum is accompanied by the removal of the previously-released Ca^{++} from the sarcoplasm. As the Ca^{++} is returned to vesicles of the sarcoplasmic reticulum, the actin and myosin filaments slide back to their former (i.e., resting) positions. At this point, the muscle cell is once again in a relaxed state and is ready to contract again when an-

other nerve impulse reaches the myoneural junction.

Motor Units

The contractions of every striated muscle fiber are regulated through their myoneural junctions. However, a single nerve fiber may give rise to branches that form myoneural junctions with a number of *different* striated muscle fibers (Fig. 4-8). The combination of *one* nerve fiber and *all* of the striated fibers that the nerve fiber regulates constitutes a **motor unit**. The ability to exercise precise control over the movements of a muscle is related to the size of its motor units. For example, in the muscles that move the eyes, each nerve fiber controls about 20 muscle cells; whereas in the muscles of the back, several hundred muscle cells are innervated by a single nerve fiber.

An entire muscle (biceps, triceps, etc.) may be thought of as a collection of motor units. The force of the muscle's contraction is determined by the number of motor units that become active. To lift a lead weight off the table by contracting the biceps muscle requires the participation of many more motor units than lifting a sheet of paper off the table.

The muscle fibers of a motor unit go through a cycle of contraction/relaxation in response to a nerve impulse, the cycle

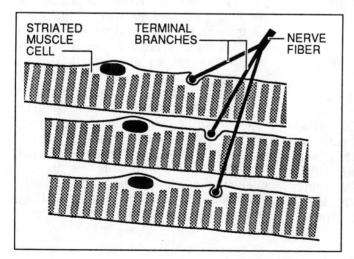

Figure 4-8
A **motor unit**. Each motor unit consists of a single nerve fiber and *all* of the striated muscle fibers that the nerve fiber stimulates. The hypothetical motor unit shown here contains three striated cells. Usually, motor units contain between 20 and 400 striated cells.

lasting less than one second. To raise an object off the table and keep it suspended requires an alternating activation of different groups of motor units. That is, the relaxation of the muscle fibers of one motor unit at the end of that unit's cycle is compensated by the activation of other, previously resting, motor units.

Isotonic vs. Isometric Contractions

The contraction of a striated muscle can take either of two forms: (1) the contraction may be **isotonic** ("iso" = "equal" and "tonic" = "strength"), or (2) the contraction may be **isometric** ("iso" = "equal" and "metric" = "length"). During an isotonic contraction, the muscle *shortens*, thereby bringing the insertion closer to the origin. An example of an isotonic contraction is the lifting of a book or some other object of similar weight off the surface of a table. In contrast, during an isometric contraction there is *no* change in the distance between the insertion and origin. An example of an isometric contraction would be the attempt to lift a 1000 pound weight off the table. The weight could not be lifted, and so the muscle's insertion would remain the same distance from the origin. This is not to imply that no contraction is occurring. Indeed, during an isometric contraction muscle fibers are shortening; however, the shortening of the muscle fibers is accompanied by an equivalent amount of stretching in the muscle's elastic tissues (e.g., in the tendons).

SMOOTH MUSCLE

As noted earlier, smooth muscle tissue is found in the walls of many of the body's internal organs. For example, in the walls of the stomach and intestine, contractions of smooth musculature mechanically degrade the food and propel it through the digestive tract; contractions of smooth muscle in the walls of the urinary bladder void the body's urine; and contractions (and relaxations) of smooth muscle tissue in the walls of blood vessels regulate the amount of blood circulating to various body parts while also regulating blood pressure.

Smooth muscle tissue is called "smooth" because the cells fail to exhibit the regular dark and light banding pattern that characterizes striated tissue. This is because the thick (myosin) and thin (actin) filaments of smooth muscle cells are not laid out in a highly ordered array (as they are in striated muscle cells). In smooth muscle cells, there are no Z-lines and thus no sarcomeres.

Structure and Arrangement of Smooth Muscle Cells

Smooth muscle cells are much shorter than striated cells and are differently shaped (Fig. 4-9). Each cell has a single nucleus that is located near the center of the cell and contributes to the bulbous shape in this region. The tapered ends of one cell overlap the bulbous regions of neighboring cells. As depicted in Figure 4-9, some cells form myoneural junctions with the terminal branches of a nerve fiber. Others, however, may have no myoneural junctions at all. Where myoneural junctions exist, these may be formed between the smooth muscle cell and the terminal branches of nerve fibers of the *sympathetic* division of the **autonomic nervous system** (see Chapter 5) or with the *parasympathetic* division. Depending on the nature of the innervation (again, see Chapter 5), the effects of the neurohumor released at the myoneural junction may be stimulatory (i.e., cause contraction) or inhibitory (i.e., cause relaxation). Cells with no myoneural junctions enter a contracted or relaxed state according to the behavior of neighboring cells in the tissue. This is because the plasma membranes of cells lacking myoneural junctions fuse with one another and also with cells that do have myoneural junctions. The points of fusion are called *gap junctions* (Fig. 4-9). Thus, polarity changes in the sarcolemma of one

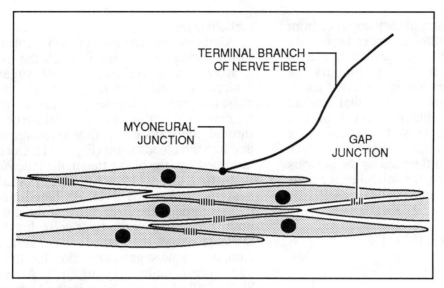

Figure 4-9
Neural innervation of smooth muscle cells (see text for details).

cell are transferred to the sarcolemma of neighboring cells .

In striated muscle tissue, contraction and relaxation may be very rapid, a complete cycle lasting less than a second. In contrast, smooth muscle tissue is characterized by slow and sustained contractions (and relaxations) that last several seconds. Other characteristics of smooth muscle tissue will be discussed in connection with the physiology of specific organ systems (e.g., the digestive system and excretory system).

CARDIAC MUSCLE

As its name implies, cardiac muscle tissue is found in the heart. Indeed, the heart is the only organ of the body to contain this special form of muscle tissue. The cells that comprise this muscle tissue (depicted in Fig. 4-10) share some physical properties common to both smooth muscle cells and striated muscle cells. Like smooth muscle cells, individual cardiac muscle cells are quite small and have only one nucleus. However, unlike smooth muscle, cardiac cells are tubular in shape and display an alternating dark and light striping pattern similar to that of striated

cells.

A distinct feature of cardiac tissue is the presence of **intercalated disks**. These are regions where finger-like projections at the ends of one cell interdigitate with the similar projections from neighboring cells. This gives added strength to the junctions between neighboring cells. Also especially notable is the absence of myoneural junctions. This is because cardiac tissue

Figure 4-10
Cardiac muscle cells reveal an A- and I-banding pattern similar to that of striated tissue. However, cardiac cells are characterized by a single nucleus and by **intercalated disks** at junctions with neighboring cells.

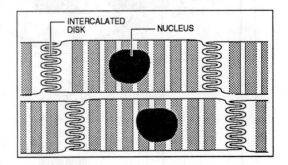

does not rely on stimulatory impulses from the nervous system in order to initiate contraction. Instead, cardiac tissue is **myogenic**; that is, the stimulus for contraction arises within the tissue itself. This is not meant to imply that cardiac tissue behaves independently of the nervous system. As you will learn in a later chapter, the rate at which the heart's cells contract is influenced by the nervous system. However, the stimulus for contraction resides within the tissue itself.

THE METABOLISM OF MUSCLE

Sugar and Glycogen

As we shall see, the actions of muscle rely heavily on the availability of carbohydrate as a fuel. Therefore, before we begin to look closely at the metabolism of muscle, it is of value to consider the body's total picture with regard to its processing of carbohydrate.

Carbohydrate present in food is broken down during its passage through the digestive tract, yielding *simple* sugar molecules. Among the simple sugars, the most common is **glucose** (also known as **dextrose**) which is then absorbed through the walls of the digestive organs and into the bloodstream (Fig. 4-11). Once absorbed into the bloodstream, the glucose is carried through the **hepatic portal system** and into the liver. As blood circulates through the liver, most of the glucose is removed by the liver cells. Inside the liver tissue, the glucose is converted from its simple sugar state into its storage form, called **glycogen**. About 80% of all of the glycogen in the body is stored in the liver; much of the remainder is stored in the body's muscles. Blood glucose that is not removed by the liver circulates to other organs of the body satisfying their fuel needs (all of the body's tissues require a continuous supply of glucose).

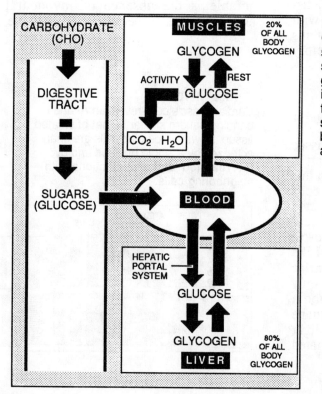

Figure 4-11
Glucose absorbed into the bloodstream from the digestive tract is stored in muscle and in the liver as glycogen. Glucose stored as glycogen in the liver may be released back into the blood on demand. In muscle tissue, glucose is broken down into carbon dioxide and water during muscle activity.

44

The conversion of glucose to glycogen is a reversible process; that is, under appropriate circumstances, glycogen can be degraded to yield individual molecules of glucose. Although the interconversion of glucose and glycogen occurs both in liver and muscle tissue, only the liver can release the sugar back into the bloodstream. That is to say, when additional sugar is needed elsewhere in the body, liver glycogen is converted back into glucose, and the glucose is then released into the bloodstream in the needed quantities. Although muscle tissue can convert glucose absorbed from the bloodstream into glycogen and use the stored glycogen in times of need, muscle tissue does not release sugar back into the blood.

Roles of Adenosine Triphosphate and Creatine Phosphate

During muscle activity, glucose is degraded in the presence of oxygen (O_2) to form carbon dioxide (CO_2) and water (H_2O). The long series of chemical reactions that brings about this process yields large quantities of energy that are used by the cell's mitochondria to form **adenosine triphosphate** (abbreviated, **ATP**) from **adenosine diphosphate** (abbreviated, **ADP**). The overall series of reactions is summarized in Figure 4-12.

ATP produced by a cell's mitochondria serves as the ultimate energy source for most other energy-requiring processes of

Figure 4-12

In muscle tissue (as well as most other tissues of the body), the energy released by the conversion of glucose to carbon dioxide and water is used to form ATP from ADP.

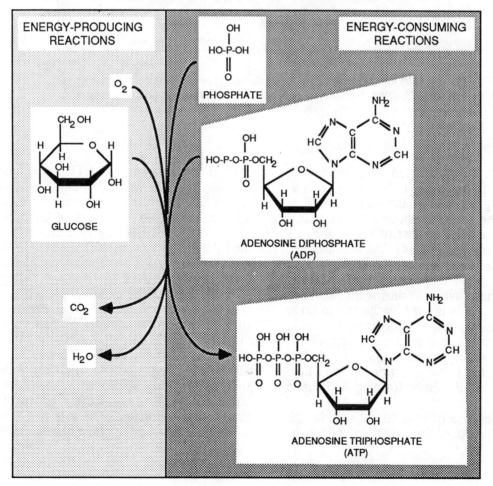

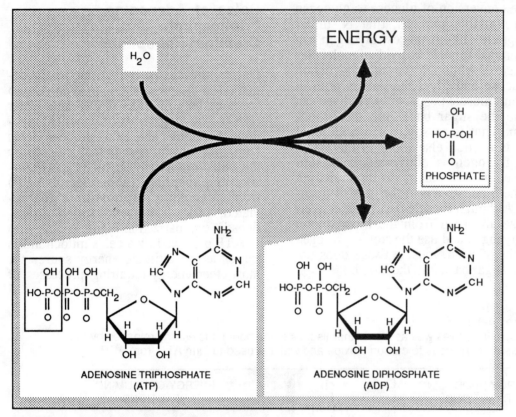

Figure 4-13
Conversion of ATP to ADP is accompanied by the release of energy that can be used by cells to drive energy-consuming metabolic reactions, such as the addition of phosphate to creatine.

the cell. For all intent and purposes, "consumption" of ATP during an energy-requiring process means that the ATP is converted to ADP; that is, the chemical bond that links the last of the molecule's three phosphate groups to the remainder of the molecule is broken, thereby releasing *free phosphate* (see Fig. 4-13). The energy that is made available during the breakage of these bonds is used by the cell to drive other energy-requiring reactions. Inside mitochondria, the terminal stages of the breakdown of glucose into CO_2 and H_2O are coupled to the addition of free phosphate to ADP thereby forming ATP again (Fig. 4-12).

In the case of muscle cells, although ATP is produced from ADP in mitochon-dria, it is sarcoplasmic ATP associated with the cell's actin and myosin filaments that is converted to ADP during filament sliding. Indeed, no movement of filaments occurs in the absence of a filament-associated ATP. Production of ATP in muscle cell mitochondria is not directly coupled to consumption of ATP by filament sliding. Instead, another chemical compound present in the sarcoplasm and called **creatine phosphate** (abbreviated **CP**) acts as an "intermediary" between mitochondrial ATP and filament ADP. That is, CP donates its phosphate to filament ADP thereby forming ATP) and uses mito-chondrial ATP to replace the phosphate that has been donated. The role of CP as an intermediary in the phosphorylation of filament-associated ADP is depicted in Figure 4-14.

The overall mechanism is summarized in Figure 4-15 and may be explained as follows. Actin and myosin filaments undergo cyclic transitions between existing in a contracted stated and being in a relaxed state. The changes between these two states are accompanied by the consumption of ATP, yielding ADP and free phosphate as products (the left side of Fig. 4-15). Glucose entering muscles cells from the bloodstream (or produced by the breakdown of glycogen within the muscle tissue [see the right side of the figure]) is converted to pyruvic acid, thereby yielding a small amount of energy that is ultimately used to form additional ATP from ADP and free phosphate. This metabolic process is called **glycolysis** and takes place in the muscle cell sarcoplasm; glycolysis does not require the presence of oxygen (i.e., the process is said to be **anaerobic**). If oxygen is available, pyruvic acid is degraded inside the cell's mitochondria to form carbon dioxide and water; the latter metabolic reactions are called **respiration**. Respiration yields large amounts of energy and is coupled to the production of large quantities of ATP from ADP and phosphate.

Creatine acts to mediate the transfer of phosphate from ATP molecules that have been produced inside mitochondria to ADP molecules that are produced during filament sliding (center of Fig. 4-15). In effect, creatine acquires its phosphate at the mitochondria (thereby forming creatine phosphate) and transfers this phosphate

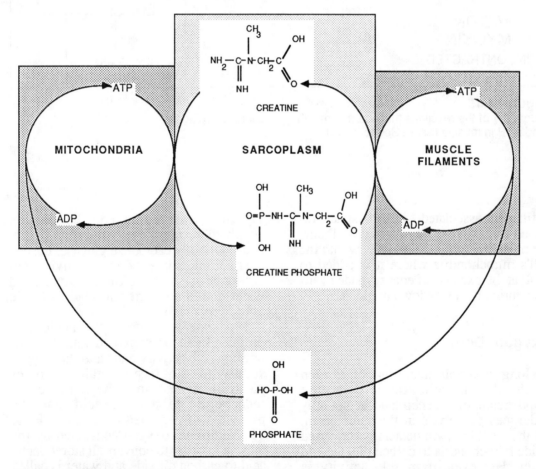

Figure 4-14
Creatine phosphate transfers phosphate to filament-associated ADP, thus replacing ATP used during muscle activity. Creatine receives phosphate from ATP produced in mitochondria. Phosphate released from ATP during filament sliding enters mitochondria and is re-incorporated into ATP.

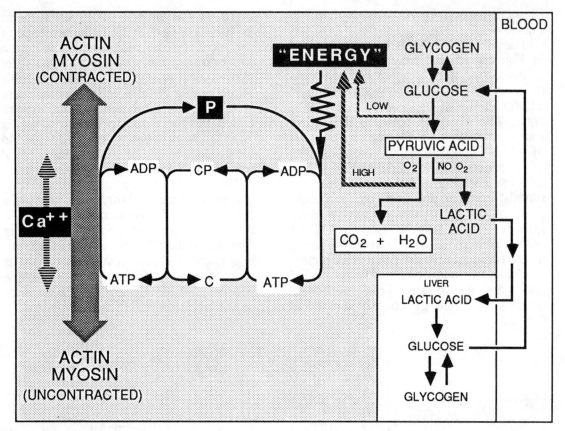

Figure 4-15
Summary of the reactions that characterize the metabolism of glucose
and ATP in muscle tissue. See text for details.

to filament-associated ADP. Phosphate released from filament-associated ATP during muscle activity makes its way into the cell's mitochondria where it is added to ADP at the expense of energy made available during the breakdown of glucose.

Oxygen Debt

So long as ample quantities of oxygen reach the muscle tissue via the bloodstream, muscle cell glucose not only undergoes glycolysis in the sarcoplasm, but the resulting pyruvic acid is converted inside mitochondria to carbon dioxide and water. However, if muscle activity is vigorous and prolonged, the oxygen supply to the muscle may fall short of that needed to fully catabolize glucose. In such a situation, some of the pyruvic acid cannot be converted to carbon dioxide and water and is converted instead to **lactic acid** (Fig. 4-15).

Moderate amounts of the lactic acid produced can be stored in muscle cells until muscle activity ends and the oxygen supply to the tissue again becomes plentiful. At that time, the lactic acid is reconverted to pyruvic acid, and the pyruvic acid degraded to CO2 and H2O. The additional oxygen consumed by the muscle in order to convert all stored lactic acid to carbon dioxide and water is called the **oxygen debt**. In other words, during vigorous activity the muscle builds up an oxygen debt, which it later repays when the activity ceases or sufficiently diminishes.

It is also possible for muscle activity to be sufficiently prolonged and vigorous that more lactic acid is formed than can be retained by the muscle cells. In this case, the excess lactic acid enters the bloodstream. Lactic acid entering the blood is carried to the liver. Here the lactic acid is removed and at considerable metabolic expense may be converted back into glucose (Fig. 4-15).

Red (Slow) vs. White (Fast) Muscle Tissue

There are two types of striated muscle tissue: red and white. The relative amounts of ATP produced in each of these tissues by glycolysis and respiration vary. The red or *slow* fibers are rich in mitochondria and employ respiration for most of their ATP production. Their demand for oxygen is satisfied in part by the presence in these cells of large quantities of an oxygen-storing protein called **myoglobin**. In contrast, white fibers contain few mitochondria and little myoglobin. In these fibers, glycolysis is the primary source of ATP. Red fibers are found in greater number in muscles with slow, rhythmic contraction patterns, whereas white fibers predominate in quickly-contracting muscles. The proportion of red to white fibers in the leg muscles of humans appears to be determined genetically and is believed to account for individual differences between long distance and sprint runners.

PHYSIOLOGY OF THE NERVOUS SYSTEM

The *nervous system* consists of all of the nerve and nerve-associated cells and tissues of the body. The main functions of this complex system are (1) to provide communication between one body part and another, (2) to interpret physical and chemical changes occurring internally, as well as external to the body, (3) to coordinate and regulate the body's activities, and (4) to store information.

The tissues of the nervous system contain a variety of cells, but one of the most highly differentiated and specialized is the **neuron**. Because many neurons are fine, thread-like cells, they are also called **nerve fibers**. The primary and specialized function of a neuron is the **conduction** and **transmission** of **impulses** from one part of the body to another. In some instances, the impulses may travel a distance of several feet, and a single neuron may bridge the entire path.

GENERALIZED ORGANIZATION OF A NEURON

The generalized structure of a neuron is shown in Figure 5-1. In certain respects, neurons are similar to most other cells. For example, the **cell body** contains the typical distribution of subcellular or-

ganelles (nucleus, mitochondria, endoplasmic reticulum, ribosomes, etc.). It is the **processes** (i.e., *extensions* from the cell body) that make neurons readily distinguishable from other types of cells. These processes are the portions of the cell that are responsible for the conduction of impulses over great (and small) distances. There are two different kinds of nerve cell processes. Processes that conduct impulses *toward* the cell body are called **dendrites**, whereas processes that conduct impulses *away* from the cell body are called **axons**. Dendrites and axons may give rise to one or more large branches or **collaterals**, each branch eventually ending in a number of fine, finger-like extensions called **terminal branches**. The terminal branches of axons are responsible either for transmitting an impulse to the next nerve cell or for innervating an "effector cell" (see later). The terminal branches of dendrites act either to receive an impulse transmitted by a nerve cell or to receive a stimulus from a "receptor cell" (again, see later). Since axons and dendrites are identified not on the basis of their lengths but on the basis of their physiological roles, the nerve cell processes labeled "axon" and "dendrite" of the nerve cell in Figure 5-1 imply that the direction of conduction in that fiber is *downward* (i.e., down the page).

Myelination of Long Processes

The long processes of many nerve cells are associated with another kind of cell called a **Schwann cell**. Schwann cells are small, sheet-like cells rich in a fatty substance called **myelin**. These cells form a *spiral envelope* around the long processes of nerve cells, thereby insulating the process from neighboring nerve cells and other tissue. Since many nerve cell processes are

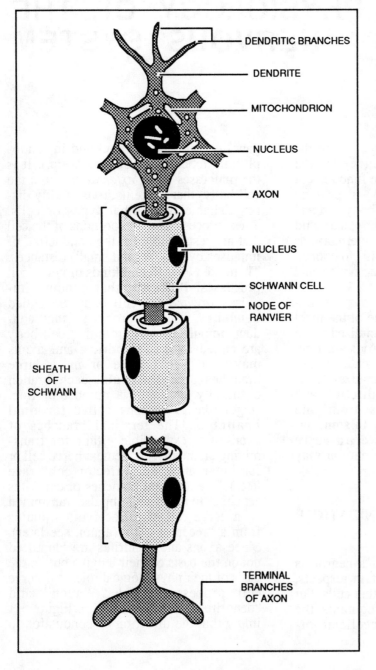

Figure 5-1
Organization of a **neuron**. Extending from the cell body are the axonal and dendritic processes. When these processes are quite long (as shown here for an axon), the process is encapsulated in a series of myelin-rich Schwann cells. The surface of the nerve cell process is exposed between successive Schwann cells; these points are called Nodes of Ranvier.

DENDRITIC BRANCHES
DENDRITE
MITOCHONDRION
NUCLEUS
AXON
NUCLEUS
SCHWANN CELL
NODE OF RANVIER
SHEATH OF SCHWANN
TERMINAL BRANCHES OF AXON

several inches (or feet) in length, the **myelin sheath** created by the Schwann cells may consist of several dozen or even hundreds of cells.

A tiny portion of the surface of a long process is exposed between successive Schwann cells. These positions are known as **Nodes of Ranvier.** Myelinated nerve cells conduct impulses more rapidly than unmyelinated cells. As you will see later, this is because the impulses quickly jump from one Node of Ranvier to the next (this is called **saltatory conduction**). Unmyelinated nerve fibers conduct impulses more slowly, the impulse passing smoothly and uninterrupted along the nerve cell surface. The myelination of processes that extend into the arms and legs and the resulting rapid saltatory conduction of impulses reduces the time taken for an impulse to travel great distances.

Figure 5-2
A simplified scheme showing the various ways in which nerve cell processes may be arranged. In *motor* fibers, the short and long processes emerge from the cell body at opposite poles of the cell. In *sensory* fibers, the short and long processes may be continuous, with the cell body linked to the long axis of the processes by a short cytoplasmic extension. In *association* fibers, several axonal and dendritic processes emerge from the cell body, and axons and dendrites may arise from neighboring positions. (Arrows indicate direction of impulse.)

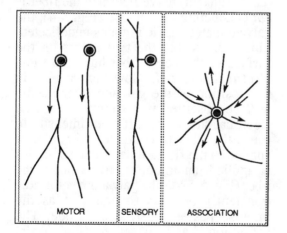

MOTOR   SENSORY   ASSOCIATION

For convenience as we continue with our discussion of the nervous system, we will use a simple, abbreviated representation of a nerve fiber. In this form, the fiber's cell body is depicted as a small circle (containing a central, dark region representing the nucleus) and the axonal and dendritic processes are shown as thin lines extending from the cell body (Fig. 5-2). Using this abbreviated style, three anatomically-distinct kinds of neurons may be distinguished: (1) **motor** neurons, in which the dendrite is usually short and the axon is usually long, the two processes exiting the cell body on opposite sides; (2) **sensory** neurons, in which the dendrite is long and the axon is short, the two processes frequently being in continuity, so that the cell body appears as a satellite; and (3) **association** neurons, in which a number of axons and dendrites of varying lengths exit the cell body at multiple points on its surface (Fig. 5-2). The physiological significance of the terms "motor," "sensory," and "association" will become apparent later.

Conduction vs. Transmission

The terms "conduction" and "transmission" have specific meanings with respect to the actions of nerve cells. **Conduction** refers to the movement of an impulse from one end of the nerve cell to the other (e.g., from the dendritic endings to the axonal endings). **Transmission**, on the other hand, refers to the passage of an impulse from the axonal endings of one nerve cell to the dendritic endings (or cell body) of the next nerve cell. Accordingly, impulses are conduced by a *single* nerve cell but are transmitted *between* cells. (It would *not* be correct to state that an impulse is "transmitted" from one end of a nerve cell to the other, or that an impulse is "conducted" from one cell to another.) As you will see later, the conduction and transmission of impulses are the results of quite different physical and biochemical phenomena

CONDUCTION OF NERVE IMPULSES

Neurons conduct impulses from one part of the body to another. Under normal circumstances each impulse begins at the dendrites (occasionally at the cell body) and spreads across the surface of the cell to the terminal branches of the cells' axons. Experimentally, an impulse can be initiated anywhere on the cell surface and can be elicited by applying a variety of stimuli including electrical shock, pressure (pinching), heat, cold, and pH changes. Once it is initiated, the impulse is propagated along the cell without dependence on a continuing stimulus. The velocity with which an impulse travels along the nerve fiber is not dependent on the strength of the stimulus; that is, it does not travel faster if initiated by a stronger stimulus.

Dromic vs. Anti-dromic Conduction

If the long process of a nerve fiber is exposed and experimentally stimulated at a point midway along its length, the fiber will simultaneously conduct an impulse away from the point of stimulation in *both* directions. One direction is the direction that leads to the terminal branches of the axon, whereas the other direction leads to the terminal branches of the dendrite. The impulse that travels in the direction of the axon's terminals is the "physiologically correct" direction (i.e., the direction in which the process normally conducts an impulse). It is called the **dromic** direction. The impulse that travels in the direction of the dendrite's terminals is "physiologically incorrect" (i.e., it is not the direction in which the cell normally conducts an impulse) and is called the **anti-dromic** direction. Anti-dromic conduction can be elicited experimentally but does not normally occur in the body. However, the ability to experimentally elicit dromic and anti-dromic conduction is a source of useful information concerning the properties and actions of nerve cells.

Nature of the Nerve Impulse

A nerve impulse is an *electro-chemical* phenomenon, but it is not the same as the conduction of electrical current in a copper wire. Electricity travels through copper wires at speeds approaching the speed of light, namely 186,000 miles per second. The movement of an impulse along a nerve fiber rarely occurs at speeds greater than about 100 miles per hour. Thus, electricity travels many hundreds of thousands times faster than a nerve impulse.

Observations Using A Galvanometer. Some of the electrical characteristics of impulse conduction may be learned from a simple experiment in which a nerve fiber is exposed and is connected to a **galvanometer**. A galvanometer is a device that measures the flow of electrical current. It consists of a meter and two measuring electrodes that can be placed on the surface of an object suspected of carrying electrical charge. For example, if the measuring electrodes of a galvanometer are attached to the positive and negative poles of a battery, the needle of the galvanometer deflects to one side, indicating that current is flowing through the galvanometer from the positive pole of the battery to the battery's negative pole (Fig. 5-3).

If the measuring electrodes of a galvanometer are placed at two distant points on the surface of an exposed (unmyelinated) nerve fiber and the fiber is *not* conducting an impulse (Fig. 5-4), the galvanometer needle remains undeflected. This implies that the two points on the surface of the nerve fiber have the same *electrical potential*. However, changes in the position of the galvanometer needle *will* be observed if a stimulus (e.g., an electrical shock) of sufficient intensity to cause the fiber to conduct an impulse is applied to the fiber. When such a stimulus is applied, an impulse travels down the fiber (Fig. 5-5A). The galvanometer needle transiently deflects to one side as the impulse passes the first electrode. The deflection that is observed (Fig. 5-5B) indicates that the surface of the fiber below the first electrode (i.e., electrode-1) is electrically negative with respect to the surface below the second electrode (i.e., electrode-2). It is as though the nerve cell surface below electrode-1 were the negative pole of a battery and the surface below electrode-2 were the positive pole of a battery. Because of the polarity differences between these two points, electrons flow through the galvanometer from electrode-1 to electrode-2.

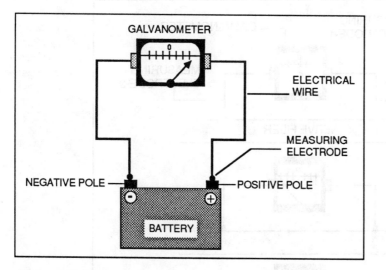

Figure 5-3
A galvanometer measures the flow of electrons between two differently-charged points. When the galvanometer's two electrodes are attached to a battery's positive and negative terminals, the galvanometer needle is deflected. The needle's deflection indicates that electrons are flowing through the galvanometer from the negative pole of the battery to the battery's positive pole. *Current* is said to flow from the positive pole to the negative pole.

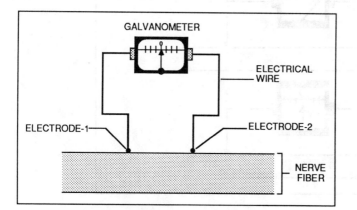

Figure 5-4
When the galvanometer's measuring electrodes are placed at two positions on the surface of a non-conducting nerve fiber, there is no deflection of the galvanometer needle. This indicates that in the non-conducting fiber, all points along the surface have the same electrical charge.

The deflection of the galvanometer needle is only temporary, because as soon as the impulse passes electrode-1, the needle returns to initial position (Fig. 5-5C). This indicates that once again the surfaces of the fiber below electrodes 1 and 2 have the same electrical potential.

As the impulse passes below electrode-2, the galvanometer needle is again deflected. However, this time the needle

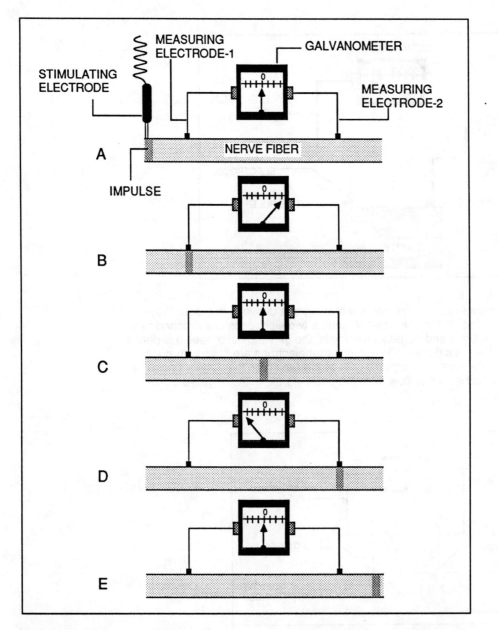

Figure 5-5
Movements of the galvanometer needle accompanying conduction of an impulse. In (A), the fiber is electrically stimulated *in front of* the first of two measuring electrodes placed on the surface of the fiber. In (B), the impulse sweeps past electrode-1. In (C), the impulse is between the two electrodes. In (D), the impulse sweeps past electrode-2. Finally, in (E), the impulse is beyond the second measuring electrode. See text for explanation of galvanometer needle movements.

moves in the *opposite* direction (Fig. 5-5D), indicating that now the surface of the fiber below electrode-2 is electrically negative relative to the surface below electrode-1 (i.e., electrons are now flowing through the galvanometer from electrode-2 to electrode-1). Finally, as the impulse proceeds beyond electrode-2, the galvanometer needle returns to its initial position. This indicates that the surfaces of the fiber below both electrodes once again have the same electrical potential.

What do these observations mean? This experiment with the galvanometer implies that (1) *the surface of a nerve fiber is electrically the same everywhere, except in the region of the nerve impulse,* and (2) *in the region representing the transient position of the impulse, the fiber's surface temporarily becomes negative relative to other regions of the surface.*

An additional informative observation may be made with a galvanometer if one of its electrodes is inserted *through* the nerve fiber's surface to the other side, while the other electrode remains on the outside surface (Fig. 5-6). When this is done, the galvanometer needle is again deflected, the deflection indicating that the inside surface of the fiber (i.e., the surface that faces the cytoplasm) is electrically negative relative to the outside surface.

Distributions of Ions Across The Nerve Fiber Surface. All of the observations described above may be explained by examining and comparing the distributions of ions across the nerve cell's surface when the fiber is "resting" (i.e., when it is not conducting an impulse) and when it is conducting an impulse. The fluid bathing the surface of a nerve fiber and the fluid within the fiber are rich in a number of ions. To make our discussion somewhat simpler, we will focus on two ions of particular importance, namely Na^+ (sodium ions) and K^+ (potassium ions). The fluid bathing the membrane's surface contains greater quantities of Na^+ than does the cytoplasm, whereas the cytoplasm contains greater quantities of K^+ than does the extracellular fluid. In addition to the Na^+ and K^+ concentration differences *across* the nerve cell membrane, the combined quantities of Na^+ and K^+ in the fluid bathing the outside surface of the membrane are greater than in the fluid bathing the inside surface. Another way of saying this is

(1) $[Na^+]_{outside} > [Na^+]_{inside}$

(2) $[K^+]_{inside} > [K^+]_{outside}$

and

(3) $\{[Na^+]_{outside} + [K^+]_{outside}\} >$

$\{[Na^+]_{inside} + [K^+]_{inside}\}$

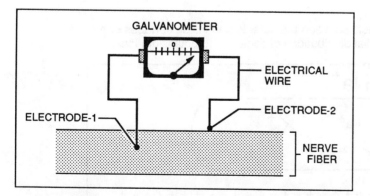

Figure 5-6
When one measuring electrode is inserted into a nerve fiber and the other electrode is placed on the fiber's surface, current flows through the galvanometer into the cell (electrons flow from the inside surface to the outside surface). This shows that the outside surface is electrically positive relative to the inside surface.

As a result of these ion distributions, the outside surface of the nerve cell membrane is positive relative to the inside surface (Fig. 5-7). This is why electrical current flows through the galvanometer when one measuring electrode is inserted through the nerve cell membrane to the other side while the other electrode remains on the outside surface of the fiber (Fig. 5-6). Because the outside surface of the fiber is electrically positive relative to the inside surface, the nerve fiber is said to be **polarized**.

The nerve cell membrane is permeable to Na^+ and K^+ and these ions are continuously diffusing from the side of the membrane where they are at higher concentration to the side where they are at lower concentration. The ions pass along their respective **concentration gradients** from one side of the membrane to the other through "pores" in the membrane; these pores are referred to as **sodium/ potassium gates**. In order to maintain the concentration differences of these two cations, some of the cell's metabolic energy is used to pump inwardly-diffusing Na^+ out of the cell and outwardly-diffusing K^+ into the cell. The mechanism that accomplishes this is known as the **sodium/potassium pump** and involves an enzyme located in the nerve cell membrane. Sodium ions together with ATP are bound to an enzyme site exposed at the inside membrane surface, and K^+ is bound to the outer surface. Binding to the

enzyme is followed by a conformational change in the enzyme that acts to move the ions through the membrane. The ATP is converted to ADP and phosphate in the process (see Chapter 4), and the sodium and potassium ions are released.

Movements of Na^+ and K^+ During Conduction. The conduction of an impulse by a nerve fiber is associated with specific changes in the distributions of Na^+ and K^+ across the nerve cell membrane. These changes are summarized in Figure 5-8. Stimulation of a nerve fiber at a point on its surface serves to **depolarize** the membrane in that region, and this widens the sodium/ potassium gates. As a result, both Na^+ and K^+ can more rapidly diffuse through the membrane along their respective concentration gradients. This effect is shown in Figure 5-8, stage 1; for purposes of easy reference, let's call this region of the nerve cell "region-1." As noted in Chapter 1, ions like Na^+ and K^+ are enclosed by layers of water molecules called **spheres of hydration**. Since the sphere of hydration around K^+ is larger than the sphere of hydration around Na^+, hydrated K^+ is much larger than hydrated Na^+. Therefore, Na^+ diffuses through the opened gates more rapidly than K^+. The inward movement of just a small amount of Na^+ in region-1 (Fig. 5-8, stage-2) is sufficient to reverse the membrane's

Figure 5-7
The difference in charge that exists between the outside and inside surfaces of a nerve cell membrane is due to differential distributions of sodium and potassium ions.

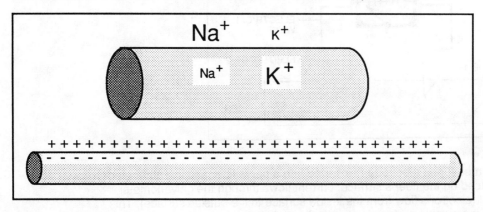

polarity in that region, thereby causing the inside surface to become electrically positive relative to the outside surface. Thus, in a localized portion of the nerve fiber (i.e., in region-1), the membrane's polarity has been reversed by these actions. The action is referred to as **reverse polarization.**

Figure 5-8
Movements of sodium and potassium ions through the nerve cell membrane during the conduction of an impulse. See text for explanation.

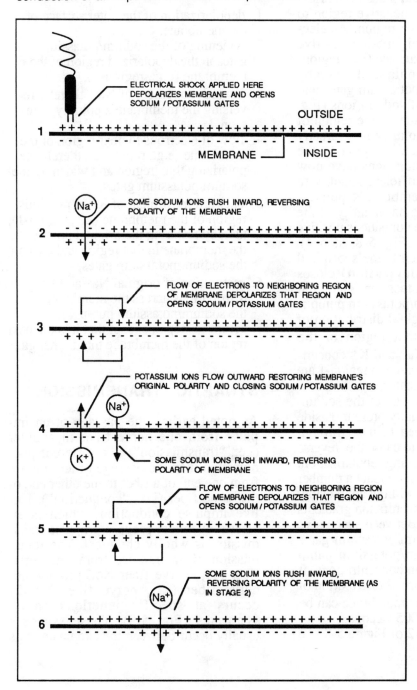

The localized reversal of the membrane's polarity creates neighboring regions on the outside (and inside) surface(s) where the electrical polarity differs. This state is depicted in Figure 5-8, stage 3, in which region-1 is negative externally and positive internally, whereas the neighboring region (let's call it "region-2") is positive externally and negative internally. This condition is unstable and leads to the flow of electrons from each negative region to the neighboring positive region. As electrons flow into the neighboring positive region on the outside surface (i.e., region-2), *that* region is depolarized, thereby widening the sodium/potassium gates and permitting an influx of sodium ions (Fig. 5-8, stage 4). Meanwhile, in the region of the fiber where the polarity had already been reversed (i.e., back in region-1), the slower-moving potassium ions have now diffused outward in sufficient quantity to again reverse the membrane's polarity (i.e., to restore the original polarity of the membrane, in which the outside is positive relative to the inside; Fig. 5-8, step 4). Restoration of the membrane's original polarity (called **repolarization**) closes down the sodium/potassium gates, at which time the sodium/potassium pump is able to restore the original distribution of Na^+ and K^+ across that region of the membrane (i.e., the Na^+ and K^+ concentration across the membrane in region-1 are returned to their initial states).

Only a small percentage of the sodium and potassium ions initially present outside and inside the cell need to pass through the membrane in order to cause the reverse polarization and the ensuing repolarization. Indeed, a nerve fiber can conduct a number of successive impulses without appreciably diminishing the ion concentration gradients across the membrane. Between successive conductions, the sodium/potassium gates narrow and the sodium/potassium pump restores the original ion concentrations. It is estimated that the normal distribution of Na^+ and K^+ across the membrane can be restored in less than 0.005 seconds.

Returning to region-2 of Figure 5-8, the influx of Na^+ has reversed the membrane's polarity in that region, again creating conditions in which the flow of electrons into the next region (say, "region-3") perpetuates the entire cycle. Thus, the conduction of an impulse may be thought of as a repeating series of events that progressively occurs along the nerve fiber's surface. The events may be summarized as follows:

1. depolarization of the outer surface of the membrane,
2. widening of the sodium/potassium gates in the depolarized region of the membrane (e.g., region-1),
3. inward movement of Na^+, thereby reversing the membrane's polarity in region-1,
4. flow of electrons to next region of the membrane (e.g., region-2), thereby depolarizing that region and widening the sodium/potassium gates,
5. outward movement of K^+ in previous region of the membrane (i.e., region-1), thereby restoring the original polarity of the membrane in that region and closing the sodium/potassium gates,
6. restoration of original Na^+ and K^+ distributions across the membrane by the sodium/potassium pump
7. return to step 3, except that it is the next region of the membrane that is changing

SYNAPTIC TRANSMISSION

As noted earlier, the passage of an impulse from one nerve fiber to another (i.e., "transmission") is a different phenomenon than the passage of an impulse from one end of a fiber to the other end of the same fiber (i.e., "conduction"). Unlike impulse conduction, which is an electro-chemical process, impulse transmission is wholly chemical. The transmission of an impulse from one nerve fiber (say, nerve fiber No.1) to another nerve fiber (say, nerve fiber No. 2) occurs at specific junctions called **synapses** (Fig. 5-9). These synapses usually occur between the axonal endings of one fiber and either the dendritic endings of the next fiber or the next fiber's cell body. The surfaces of the fibers at a synapse are separated from one-another by a narrow gap called the **synaptic cleft.**

Transmission of an impulse from one nerve fiber to another generally takes the following form. The arrival of the impulse at the terminal branches of fiber No. 1 is followed by the release from those branches of a chemical substance called a **neurotransmitter** or **neurohumor**. The most common of the body's neurotransmitters is **acetylcholine**; other less common neurohumors include **norepinephrine, dopamine**, and **seratonin**. The neurohumor released into the synaptic junction diffuses across the small gap that separates the surfaces of the two fibers and binds to transmitter-receptors in the surface of fiber No. 2. The effect of neurohumor binding is to widen the sodium/potassium gates in that region of the membrane and in so doing to trigger the events of conduction in fiber No. 2, as described in detail above. The process is repeated at the next synapse.

In the simplest case, the transmission of an impulse involves only two fibers, namely a "pre-synaptic" fiber and a "post-synaptic" fiber. However, several fibers may form a synapse, so that signals may pass from the axonal endings of one or more of the fibers to the dendritic endings of other fibers forming the synapse. Moreover, several *different* neurohumors may be released by different cells forming a synapse, and the neurohumor released by one cell may act to *block* the transmission of an impulse that would otherwise be achieved through the action of a neurohumor released by another cell.

In the case of acetylcholine, acetylcholine that is bound to acetylcholine receptors of the plasma membrane of the post-synaptic nerve cell is broken down into *choline* and *acetate* by the enzyme **acetylcholinesterase**. This enzyme is located on the surface of the post-synaptic fiber. The acetate and choline are then released into the synaptic cleft. Shortly afterwards, the acetate and choline are taken up by the pre-synaptic fiber, which recycles these substances so that new molecules of acetylcholine are formed.

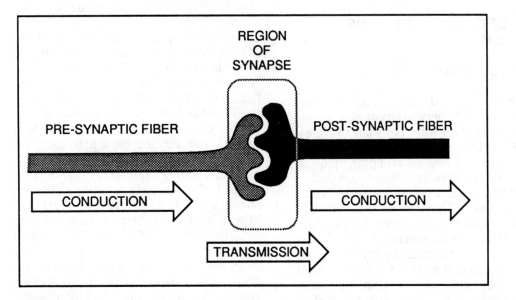

Figure 5-9
A **synapse** is a junction between two nerve fibers. At this junction a nerve impulse is **transmitted** from the pre-synaptic fiber to the post-synaptic fiber.

MAJOR SUBDIVISIONS OF THE NERVOUS SYSTEM

The nervous syetm may be subdivided into 3 major portions: (1) the **central nervous system** (or **CNS**), (2) the **peripheral nervous system** (or **PNS**), and (3) the **autonomic nervous system** (or **ANS**).

The central nervous system consists of the **brain** and the **spinal cord**. The peripheral nervous system consists of the nerves that arise from the brain and the spinal cord, namely the **cranial nerves** (arising from the brain) and the **spinal nerves** (arising from the spinal cord). The autonomic nervous system includes a number of nerves that regulate the activities of the body's internal organs (e.g., heart, digestive organs, and excretory organs). The autonomic nervous system has two divisions of its own; these are called the **sympathetic** division and the **parasympathetic** division. We will proceed with our discussion of the nervous system by considering the system's divisions in the order that has just been presented.

THE CENTRAL NERVOUS SYSTEM

Grey vs. White Matter

Before we proceed any further into our discussion of the nervous system, it is important to explain what is meant by "grey matter" and "white matter," since both of these terms will now be cropping up quite regularly. The myelin that forms a sheath around many nerve cell processes is white in appearance. Thus, nerve tissue that is rich in myelinated processes looks white and is called **white matter**. Even when a nerve cell is myelinated, the myelin sheath does not extend over the cell body, which has a grey appearance. Consequently, nerve tissue that is rich in nerve cell bodies and unmyelinated processes appears grey and is called **grey matter**.

The Brain

The brain of the average adult contains about 100,000,000,000 (i.e., 100 billion!) nerve cells, virtually all of which are produced during embryonic development. (Very few new nerve cells are produced after birth so that any nerve cells lost through disease or injury are not replaced.) There are three anatomically and functionally different parts of the brain. The main part of the brain is the **cerebrum** (Fig. 5-10), which is further divided into left and right halves (i.e., the left and right *cerebral hemispheres*).

Underneath the rear of the cerebrum is the **cerebellum**; connecting the cerebrum to the cerebellum and also linking both of these to the top of the spinal cord is the **brain stem**. Supporting the cerebrum and cerebellum from below and also encapsulating the brain at its surface (thereby physically protecting the brain) is a thick sheet of bone called the **skull**. The **cranial nerves** (see below) emerge through small openings in the skull.

The cerebrum is characterized by the many grooves that run across its surface. The deeper grooves are called **fissures** and the elevations between neighboring fissures are called **convolutions**. The fissures give rise to shallower grooves called **sulci**, which therefore subdivide the convolutions into a number of neighboring **gyri**. The deepest of the fissures separate each of the two cerebral hemispheres the into 4 different lobes: (1) the **frontal** lobe, (2) the **parietal** lobe, (3) the **occipital** lobe, and (4) the **temporal** lobe. For the most part, each lobe of the cerebrum regulates the activities of a particular region of the body (although the control of some body functions clearly is shared by two or more lobes). If one cuts into the surface of the cerebrum, one notes that the outer layer (called the **cerebral cortex**) is grey, indicating that it is comprised predominantly of nerve cell bodies and unmyelinated processes. Below the cortex the cerebral tissue is white, indicating the presence of many myelinated processes.

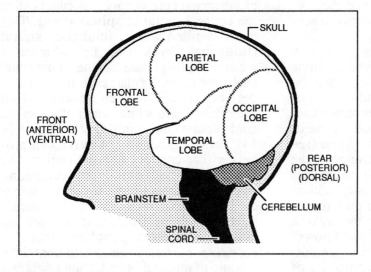

Figure 5-10
The brain is comprised of three major parts; these are the **cerebrum** (which is divided into four lobes), the **cerebellum**, and the **brainstem**. Seen in this diagram is the *left* side of the brain.

Each cerebral lobe contains three functional regions called the **sensory, motor**, and **association** areas. The sensory area of a cerebral lobe is the area that receives signals from distant parts of the body and is believed to be the area in which the incoming signals are *interpreted*. Thus, "understanding" is a phenomenon often ascribed to the sensory areas of the cerebral lobes. The motor area of a lobe is the region from which signals emerge; these signals are carried to distant parts of the body where they bring about some form of body action. The association area of a lobe serves to link the sensory and motor areas and is believed to be the region in which information is stored (memory) and in which judgements are rendered. The association areas are also though to be the source of emotions. As we delve further into human physiology, we will consider the actions of the cerebrum in greater detail.

As noted above, the cerebellum is a portion of the brain located beneath the rear of the cerebrum (i.e., below the occipital lobes; Fig. 5-10). The principal function of the cerebellum is the subconscious regulation of muscle activity in such commonplace body actions as standing upright, walking, and running. The cerebellum is in continuous communication with the muscles and joints of the body, receiving sig-

nals from these remote regions as the muscles contract, relax, or are stretched. In return, the cerebellum sends signals to the musculature in order to ensure smooth and coordinated muscle activity.

Although influenced by the overlying cerebrum, the brain stem (Fig. 5-10) serves to subconsciously regulate and coordinate the body's **vegetative functions**, including *heart rate*, *digestion*, and *respiration*. The regulatory signals originate in a number of "centers" (e.g., *cardioaccelerator center*, *inspiratory center*, etc.) located in portions of the brainstem known as the **midbrain, pons**, and **medulla**.

The Spinal Cord

Beginning at the base of the brainstem and descending several feet vertically is the rod-like **spinal cord**. Unlike the brain, which is enclosed and protected by a *continuous* layer of bone (i.e., the skull), the spinal cord is protected by a series of bony rings called **vertebrae**. The vertebrae collectively form the **vertebral column**. Successive vertebrae are separated from one another by pads of cartilage called **disks**; this arrangement of vertebrae and disks provides protection while also allowing the spinal cord to bend or twist according to body movement and body position. The **spinal nerves** (see below) emerge from the spinal cord between successive vertebrae (Fig. 5-11).

Figure 5-12 shows a cross-section through the spinal cord. As you examine this figure, it is important to note that the *inner* portion of the spinal cord consists of grey matter, whereas the *outer* region consists of white matter. This is the reverse of the condition that exists in the brain (see above). The basis for the differences in the relative positions of white and grey matter in the brain and in the spinal cord is illustrated in Figure 5-13. As the grey matter passes up the spinal cord into the brain, it extends to the brain's surface and "flows" over and around the brain; thus, the outside surface of the brain is composed of grey matter–not white matter.

Returning to Figure 5-12, at the center of the spinal cord is a narrow canal called the **central canal** or **spinal canal**. This canal is filled with a fluid (i.e., **spinal fluid**) that serves in the nourishment of the surrounding tissue. As the spinal canal enters the brain, it enlarges to form a number of fluid-filled chambers called "ventricles." The white matter of the spinal cord contains the **spinal tracts** (Figure 5-14); these are bundles of nerve cell processes that conduct impulses up and/or down the spinal cord. Tracts containing fibers that conduct impulses *up* the spinal cord are called *ascending tracts*, whereas tracts containing fibers that conduct impulses *down* the spinal cord are called *descending tracts*. Some tracts contain a mixture of upwardly-conducting fibers and downwardly-conducting fibers; these are called *mixed tracts*.

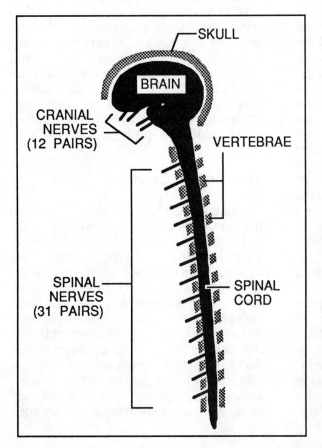

Figure 5-11
The brain and spinal cord comprise the **central nervous system**. Emerging from the brain (through tiny apertures in the skull) are the 12 pairs of **cranial nerves**. Emerging from the spinal cord between successive vertebrae are the 31 pairs of **spinal nerves**. The cranial and spinal nerves comprise the **peripheral nervous system**.

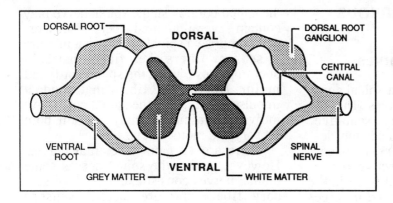

Figure 5-12
A cross-section through the spinal cord and the dorsal and ventral roots of a pair of spinal nerves. Note that the spinal cord's grey matter is enclosed by the white matter.

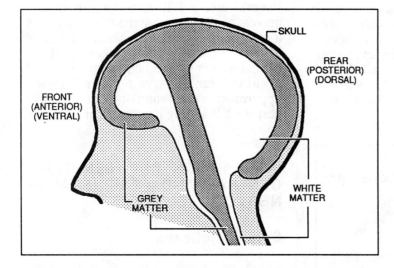

Figure 5-13
Whereas the grey matter of the spinal cord is surrounded by white matter, the brain's grey matter is at the surface. This diagram illustrates how this reversal occurs. See the text for an explanation.

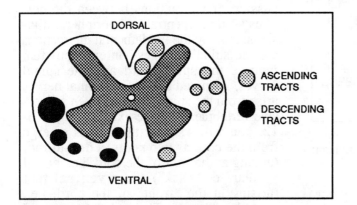

Figure 5-14
The nerve cell processes that conduct impulses up or down the spinal cord are gathered together into bundles called **spinal tracts**. The tracts are in the spinal cord's white matter.

THE PERIPHERAL NERVOUS SYSTEM

Cranial and Spinal Nerves

Emerging from the brain and spinal cord are the cranial and spinal nerves. The internal organization of a representative nerve is shown in the cross-section depicted in Figure 5-15. The nerve is enclosed by a band of connective tissue called **epineurium**. Internally, the individual nerve fibers are arranged in large bundles called **fasciculi** (singular = **fasciculus**). (The fasciculi of a cranial or spinal nerve are analogous to the tracts of the spinal cord.) Each fasciculus is covered by a layer of connective tissue called **perineurium**. Within each fasciculus, groups of nerve fibers are embedded in yet another form of connective tissue called **endoneurium**. The endoneurium is separate and distinct from the myelin sheaths that may envelope specific fibers.

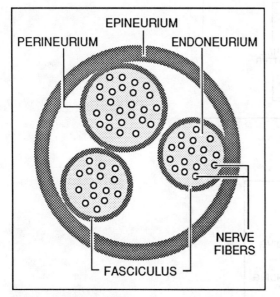

Figure 5-15
A cranial or spinal nerve seen in cross-section; see text for details.

There are 31 pairs of spinal nerve, one member of each pair exiting from the right half of the spinal cord and the other member of the pair exiting from the left half of the spinal cord. A single spinal nerve may contain several hundred fibers. The fibers that comprise a spinal nerve are a mixture of **sensory fibers** (i.e., fibers that conduct impulses *toward* the spinal cord) and **motor fibers** (i.e., fibers that conduct impulses *away from* the spinal cord). An individual fasciculus may contain both sensory and motor fibers.

There are 12 pairs of cranial nerves. However, unlike the spinal nerves some cranial nerves consist exclusively of sensory fibers and are called "sensory nerves." The remaining cranial nerves, like all spinal nerves, are "mixed nerves" (i.e., they contain both sensory and motor fibers).Table 5-1 lists the various cranial nerves; as you examine the table, note that it is customary to idenitfy the nerves using Roman numerals. The numbering of the cranial nerves is based on their anatomical position, cranial nerve number I arising uppermost in the brain, and cranial nerve number XII arising lowermost in the brain.

INTERACTION BETWEEN THE CENTRAL AND PERIPHERAL NERVOUS SYSTEMS

Spinal Reflexes

Using the forgoing description of the organization of the nervous system as a foundation, we may now proceed to a discussion of interactions between the nervous system's central and peripheral divisions. Among the fundamental interactions are the spinal reflexes—interactions that involve the spinal cord and the spinal nerves. As noted earlier, the spinal nerves arise from the spinal cord between successive vertebrae. If you re-examine Figure 5-12, you will see that the nerves emerge from the cord as two **roots**: a **dorsal** root (arising at the *rear* of the cord; i.e., dorsal = "rear" or "back") and a **ventral** root (arising at the "front" of the cord; i.e., ventral = "front"). The dorsal and ventral roots of each nerve merge a short distance from the spinal cord.

TABLE 5-1 THE CRANIAL NERVES

NUMBER	NAME	TYPE	FUNCTIONS
I	Olfactory	Sensory	Olfaction (sense of smell)
II	Optic	Sensory	Vision
III	Occulomotor	Mixed	Movements of the eyes; accommodation of the lens; pupil size; proprioception
IV	Trochlear	Mixed	Eye movements; proprioception
V	Trigeminal	Mixed	Facial sensations; chewing
VI	Abducens	Mixed	Eye movements; proprioception
VII	Facialis	Mixed	Facial expression; gustation (sense of taste)
VIII	Acoustic	Sensory	Sense of hearing; equilibrium
IX	Glossopharyngeal	Mixed	Swallowing; salivation; gustation
X	Vagus	Mixed	Sensations from and motor control over many internal organs (e.g., heart)
XI	Accessory	Mixed	Head movement; proprioception
XII	Hypoglossal	Mixed	Speech; proprioception

Simple Spinal Reflexes

During a physical examination, it is not uncommon for the physician to perform a test in which you are asked to cross your legs, following which you are gently struck below the knee with a rubber mallet. The normal response is a sudden (and involuntary) extension (or "jerk") of the crossed leg. This reaction is an example of a **simple spinal reflex**. "Postural reflexes" are another illustration of simple spinal reflexes. In a postural reflex, contractions of the musculature of the back and/or neck maintain the body and head in an upright position. For example, you may have experienced a situation in which you were seated upright watching television or listening to a lecture and began to drift off to sleep. As your head started to tilt sideways or forward, the postural reflex suddenly brought your head back to its upright position, perhaps startling you in the pro-

cess. These and other simple spinal reflexes are founded on the following interactions.

The typical simple spinal reflex (Fig. 5-16) begins when **receptor cells** sense physical or chemical changes within the body or near to the body surface. For example, in the knee jerk, receptor cells sensitive to pressure changes in the skin and underlying muscles respond to the compression of the tissues below the knee. Receptor cells are associated with the dendritic nerve endings of **sensory nerve fibers**, so that activation of the receptors is quickly followed by stimulation of the dendritic endings. The result is the propagation of a nerve impulse along the sensory fibers.

The dendrites of the sensory fibers become part of a spinal nerve. Each fiber's cell body is housed in a small bulbous enlargement of the spinal nerve's dorsal root. This enlargement is called a **dorsal root**

ganglion. The axonal process of each sensory nerve fiber extends from the dorsal root ganglion into the grey matter of the spinal cord. Within the spinal cord's grey matter, the axonal processes of the sensory fibers synapse with either the dendritic endings or the cell body of **motor** nerve fibers. The nerve impulses conducted along the sensory fibers are transmitted to the motor fibers in the grey matter. The axonal processes of the motor fibers exit the grey and white matter of the spinal cord through the ventral root and become a part of the spinal nerve. The motor fibers extend into the leg musculature and branch into a number of fine endings that form junctions with striated muscle fibers. Impulses reaching these striated cells cause contraction. The arrangements of a receptor, sensory fiber, motor fiber, and **effector** (i.e., muscle cell) are depicted in Figure 5-16).

There are several important points to be kept in mind regarding a simple spinal reflex and its illustration in Figure 5-16. First of all, note that there are only two types of nerve fibers involved in a simple spinal reflex: sensory fibers and motor fibers. However, this is not intended to imply that the completion of a simple spinal reflex employs only two fibers. Indeed, depending upon the intensity with which the mallet strikes the leg, several hundred sensory fibers may be caused to conduct impulses into the spinal cord. By

the same token, several hundred motor fibers may conduct impulses out of the spinal cord and on toward several hundred striated cells in the leg muscles that jerk the leg upwards. Finally, it is to be emphasized that the reflex is mediated through the spinal cord and does not require participation of the brain in order to be completed (i.e., a knee jerk reflex can be elicited from an unconscious person). This does not mean that one is unaware of the impact of the mallet, it simply means that the reflexive action occurs whether or not the brain is "aware" of what has taken place.

OTHER SPINAL REFLEXES

We will now turn to other spinal reflexes that are somewhat more complex. A good illustration is the sudden reflexive withdrawal of one's hand from the surface of a hot object that is inadvertently touched. The pathways involved in this reflex are shown in Figure 5-17.

Again, the reflex begins with the stimulation of receptors. In this case, the receptors are cells in the skin that are sensitive to heat. The stimulated receptors respond by triggering nerve impulses in the sensory fibers with which the receptors are coupled. As in a simple spinal reflex, the impulses are conducted into the grey matter of the spinal cord through the spinal nerve's dorsal root.

Figure 5-16
Pathway of a *simple spinal reflex*. See text for discussion.

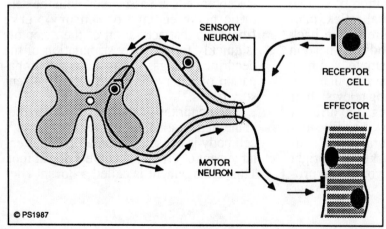

© PS1987

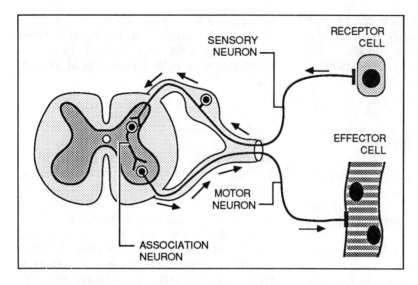

Figure 5-17
Pathway of a *non-simple* spinal reflex. See text for discussion.

Within the grey matter of the spinal cord, the impulses are transmitted to nerve fibers called **association** fibers. Some of these fibers then conducts impulses to the dendritic endings (or cell bodies) of motor fibers; the latter complete the reflex by conducting impulses to the effector cells (e.g., muscle cells in the arm that cause you to lift your hand off the hot object). As in a simple spinal reflex, the motor impulses are conducted from the spinal cord via the ventral root. Notice that a major difference between this reflex and the one described earlier is the involvement of a third class of nerve fibers, namely association fibers.

The interposition of association fibers between sensory fibers and motor fibers has a number of consequences. One of these is the ability of the brain to easily override the reflex. For example, suppose that you place your hand on an object that is hot but is not so hot that it could injure (i.e., burn) the skin. As the skin receptors respond to the sudden rise in temperature, the spinal reflex is triggered and you begin to withdraw your hand. However, sensing that the object is not too hot to touch, you can return your hand to the warm surface

and keep it there. In such a scenario, what you are doing is "overriding" the spinal reflex. This is achieved physiologically by sending nerve impulses down through tracts in the white matter of the spinal cord from the brain to the junctions between association and motor fibers of the reflex arc. These impulses serve to inhibit the transmission of signals from association to motor fibers and therefore prevent you from pulling your hand away from the hot object. In other words, you can exercise conscious control over this kind of spinal reflex.

Flexion and Extension

The two reflexes described above illustrate two alternative classes of muscle actions: **flexion** and **extension**. By flexion is meant contraction of a muscle that serves to move a body part (usually an arm or leg) closer to the **torso** (e.g., closer to the head, chest, or abdomen). For example, when you pull your hand away from the surface of a hot (or sharp) object, you pull your hand toward the body (i.e., you don't simply move it to one side or the other).

By extension is meant a muscle contraction that serves to move a body part further away from the torso. This is illustrated by the knee jerk, in which muscle contraction extends the lower leg, so that the foot is further away from the torso.

Reciprocal Innervation

Flexion of a muscle is accompanied by stretching of the *antagonistic* muscle. For example, when the biceps muscle of the upper arm contracts, the arm is flexed (i.e., the lower arm is raised toward the shoulder). As a result, the triceps muscle of the upper arm is stretched. By the same token, extension of the lower arm brought about by contraction of the triceps muscle serves to stretch the biceps muscle. Because they have opposite effects, the biceps and triceps muscles are said to be antagonistic. Clearly, it would be counterproductive if a pair of antagonistic muscles were to contract *at the same time*. When one of a set of antagonistic muscles is stimulated by motor pathways and caused to contract, the antagonistic muscle is in-hibited via a separate set of motor pathways and thereby is prevented from contracting. This important physiological mechanism is called **reciprocal inner-vation**.

The Crossed-Extensor Reflex

An interesting spinal reflex that involves both flexion and extension is the **crossed-extensor reflex**. This reflex may be illustrated as follows. Suppose that a person walking barefoot steps on a thumb-tack with his left foot. Puncturing of the skin of the foot and injury to underlying receptors triggers a reflex in which there is flexion of the left leg as flexor muscles contract and extensor muscles are relaxed by reciprocal innervation. This raises the left foot off the ground, and by taking the weight of the sole of the foot prevents further penetration of the thumbtack. At the same time, there is extension of the right leg as extensor muscles contract and allow the body weight to be supported by one leg. The nerve pathways involved in this reflex are shown in Figure 5-18.

Figure 5-18
The **crossed-extensor reflex**. This spinal reflex involves **reciprocal innervation** of the muscles of the legs and invokes the **flexion** of one leg and **extension** of the other. See text for discussion.

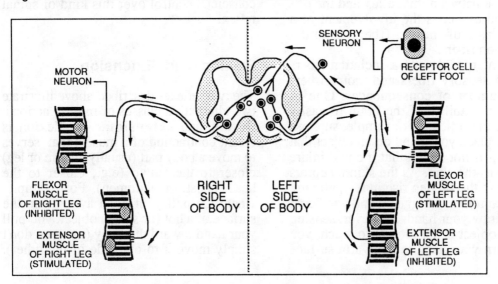

The reflexive actions of the left leg are said to be **ipsilateral**, which means that the incoming sensory signals (from the sole of the left foot) and outgoing motor signals (to the flexor and extensor muscles of the left leg) involve only one side of the body and one side of the nervous system (i.e., the left side). In contrast, the reflexive actions of the right leg are said to be **contralateral** (i.e., they involve the opposite side of the body and opposite side of the nervous system).

In all of the reflexes that we have so far considered, the incoming sensory impulses and the outgoing motor impulses involve the same level of the spinal cord. For example, the knee jerk reflex involves only the pelvic region of the spinal cord, and withdrawing one's hand from a hot object involves only the shoulder level of the spinal cord.

Some reflexive actions, however, involve sensory input and motor output at different levels of the spinal cord. For example, when you are abruptly pushed on the shoulder, displacing your body to one side, you reflexively extend the leg on the opposite side of the body in order to avoid falling. (Note that the above action is contralateral.) Whenever either contralateral or ipsilateral reflexes involve two (or more) levels of the spinal cord, the conduction of impulses between the different levels is mediated by association fibers whose long processes are part of a spinal tract.

THE AUTONOMIC NERVOUS SYSTEM

The autonomic nervous system includes a variety of motor pathways that control the actions of many of the body's internal organs. These motor pathways are divided anatomically and physiologically into two groups: (1) pathways of the **sympathetic division**, and (2) pathways of the the **parasympathetic division**. It is, perhaps, easiest to understand the overall actions of these two divisions by considering examples of what each division does. Let us begin by listing some of the physiological changes that occur in the body when the sympathetic division of the autonomic nervous system becomes active.

Actions of the Sympathetic Division

1. The amount of light entering each eye is regulated by a smooth muscle called the **iris**. The iris muscle is shaped like a donut, the opening at the center forming the eye's **pupil** (i.e., the central black spot); the iris musculature also contains pigments that determine one's eye color (i.e., brown eyes, blue eyes, etc.). The muscle tissue of the iris is arranged in two ways: circumferentially and radially. Contraction of the circumferential muscle tissue makes the pupil smaller, reducing the amount of light entering the eye; contraction of the radial muscle tissue dilates the pupil, allowing more light to enter the eye. Motor fibers of the sympathetic division innervate the radial muscle tissue of the iris, thereby causing dilation of the pupil.

2. Motor fibers of the sympathetic division also innervate the **salivary glands** (glands in the walls of the mouth that secrete saliva). The effect of impulses reaching the glands over these sympathetic pathways is to *inhibit* secretion, causing the mouth to become dry.

3. Motor fibers of the sympathetic division also innervate the **sweat glands** (glands in the skin that secrete a watery fluid onto the skin's surface, thereby cooling the skin). The effect of impulses reaching the sweat glands over these sympathetic pathways is *stimulatory*, thereby causing perspiration.

4. The walls of the digestive organs (stomach, small intestine, etc.) are rich in smooth musculature, the contractions of which help to digest the food and propel it through the digestive tract. The

musculature of the digestive tract receives nerve impulses over autonomic pathways. In the case of the sympathetic division, the effect on this musculature is *inhibitory*, bringing a halt to such digestive activity.

5. The walls of the digestive organs are also rich in blood vessels and the walls of these vessels contain smooth muscle tissue. The role of this muscle tissue is to regulate the diameter of the blood vessel, decreasing the diameter by contracting and increasing the diameter by relaxing. Increasing the vessel's diameter increases the flow of blood through the vessel and decreasing the diameter reduces the flow of blood through the vessel. Smooth muscle tissue in the walls of these blood vessels is innervated by the autonomic nervous system, and the effect of the sympathetic division is to cause *contraction*. This reduces the diameter of the vessels, thereby reducing the flow of blood to the digestive tract. (Note that this action is consonant with number 4, above.)

6. The autonomic nervous system also regulates the flow of blood to the skeletal musculature by either dilating or constricting the blood vessels that carry blood to these organs. In the case of the sympathetic division, the action is to *inhibit* smooth muscle cells in these blood vessels so that the vessels dilate, thereby conducting more blood into the skeletal musculature. Thus, blood that is diverted from the digestive organs by the action of the sympathetic division (see number 5, above) is carried instead to the skeletal muscles.

7. The passageways that carry air to the lungs (i.e., the **bronchi** and **bronchioles**) have smooth muscle tissue in their walls. Motor impulses reaching this muscle tissue via fibers of the sympathetic division *inhibit* these cells, thereby causing dilation of the air passageways.

8. Finally, the control centers of the heart (i.e., the **sino-atrial node** and **atrio-ventricular node**) also receive impulses from the autonomic nervous system. The effect of the sympathetic division is *stimulatory*, causing an increase in heart rate.

By considering these eight effects together, it is clear that the sympathetic division of the autonomic nervous system can bring about major changes in the body's physiological state: the pupils dilate, the mouth becomes dry, the skin perspires, digestion of food ceases, additional blood flows to the skeletal muscles, large amounts of air are readily conveyed to the lungs, and the heart begins to beat faster. All of these events serve to switch the body's level of physiological and metabolic activity from a normal or "idling" state to one in which there can be a sudden and massive expenditure of energy. The change is often referred to as the **"flight or fight reaction,"** in the sense that the body can either quickly flee from a threatening situation or "take a stand" and expend the energy in combat. Of course, some or all of these physiological changes may take place in the absence of a genuine threat (such as watching a frightening film), but it is the action of the motor fibers of the sympathetic division that bring about the physiological changes.

Actions of the Parasympathetic Division

As you might suspect at this point, the action of the parasympathetic division of the autonomic nervous system is to return the body from a high level of metabolic and physiological activity to a resting or normal state. The parasympathetic division's motor fibers conduct impulses to the same effectors as the sympathetic division's fibers; however, *the effect on the target tissue is just the opposite*. Whereas the sympathetic division dilates the pupils, the parasympathetic divsion constricts the pupils; whereas, the sympathetic division increases heart rate, the parasympathetic divsion reduces heart rate; and so on. Indeed, any effect of the sympathetic division may be reversed by the parasympathetic division. Acting cooperatively, the sympathetic and parasympathetic divisions of the autonomic nervous system adjust the levels of activity of the various organs in order to meet the body's prevailing needs .

Organization of the Sympathetic Division

Let us now turn to a consideration of the manner in which the motor fibers of each division of the autonomic nervous system are arranged and organized. Unlike the motor pathways that we have considered previously (in which a single motor neuron extends the entire distance from the central nervous system to the effector), motor pathways of the autonomic nervous system always consist of *two* successive fibers. In the case of the sympathetic division, the cell body of the first fiber lies in the grey matter of the spinal cord. Its axonal process exists the spinal cord through the ventral root of a spinal nerve. The process soon leaves the nerve and enters the nearest **lateral ganglion** in a chain of 22 **sympathetic ganglia** that lie one on

each side of the vertebral column in the thoracic and lumber regions of the body (Figure 5-19). The first fiber may synapse in that ganglion with the second fiber of that pathway or it may ascend or descend to another ganglion in the chain before synapsing. It may even pass through the ganglion of the sympathetic chain and synapse in one of several **collateral ganglia** that lie several inches away from the sympathetic chain (Figure 5-20). The first of the two motor fibers is called the **pre-ganglionic fiber**, whereas the second fiber of the pathway is called the **post-ganglionic fiber**. The axonal process of the post-ganglionic fiber terminates at its junction with the effector.

Organization of the Parasympathetic Division

Like the sympathetic division, the motor pathways of the parasympathetic division consist of two successive fibers (a pre-ganglionic fiber and a post-ganglionic fiber). The pre-ganglionic fibers emerge from two different regions of the central nervous system: the brain and the bottom tip of the spinal cord. Those that emerge from the brain enter certain cranial nerves (most enter the Vagus nerve).

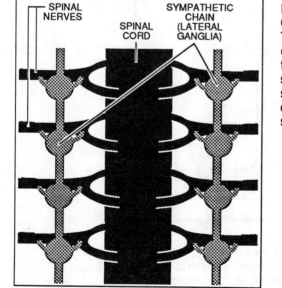

Figure 5-19
Organization of the sympathetic chains. The sympathetic chains are two chains of interconnected ganglia that descend through the body on each side of the spinal cord, just ventral to the emerging spinal nerves. The ganglia are also connected to the ventral roots of the spinal nerves.

73

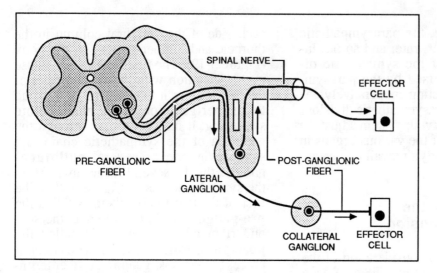

Figure 5-20
In a sympathetic pathway, the **pre-ganglionic fiber** exits the spinal cord through a ventral root and later synapses with a **post-ganglionic fiber** either in a **lateral ganglion** or in a **collateral ganglion**.

Those pre-ganglionic fibers that originate in the grey matter at the bottom of the spinal cord exit the cord through the ventral root of a spinal nerve. In the parasympathetic division, the ganglia in which pre- and post-ganglionic fibers synapse lie close to the organs that are regulated Figure 5-21). Thus, pre-ganglionic fibers are long, whereas post-ganglionic fibers are very short. The ganglia in which synapsis occurs are called **terminal ganglia.**

As noted above, the autonomic nervous system regulates the activities of many of the body's internal organs. The relationship between the sympathetic and parasympathetic divisions and their pre- and post-ganglionic fibers is illustrated for several organs in Figure 5-22. For simplicity, only 5 lateral and 2 collateral ganglia are shown.

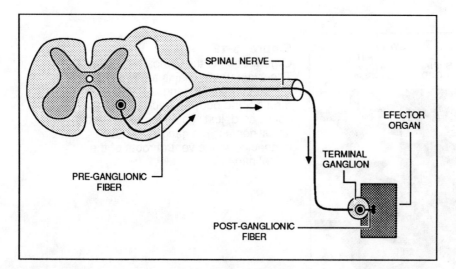

Figure 5-21
In a parasympathetic pathway, the pre-ganglionic fiber exits the spinal cord through a ventral root and later synapses with a post-ganglionic fiber in a **terminal ganglion** located in or near the surface of the effector.

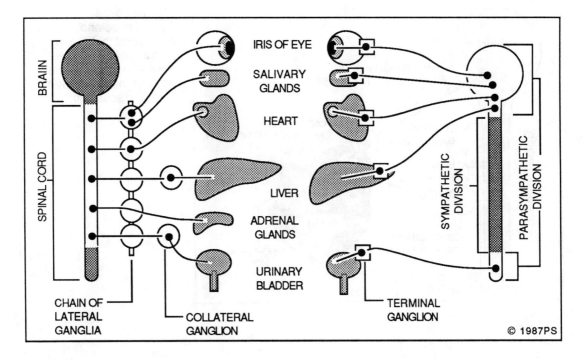

Figure 5-22
A comparison of the motor pathways (pre-ganglionic fibers, post-ganglionic fibers, and ganglia) of the sympathetic and parasympathetic divisions of the autonomic nervous system.

Neurohumors of the Autonomic Fibers

In both divisions of the autonomic system, the axonal endings of pre-ganglionic fibers release the neurotransmitter acetylcholine (i.e., the same neurohumor discussed earlier in the chapter). Acetylcholine is also released at the junctions between the post-ganglionic fibers of the parasympathetic division and the effector (e.g., at the junction with muscle or gland tissue). However, in the sympathetic division, most post-ganglionic endings release the neurotransmitter **nor-epinephrine**. Since the sympathetic and parasympathetic divisions of the autonomic nervous system have opposite effects on the tissues and organs with which they communicate, it is not surprising that the neurohumors re-

leased at the effector junctions are different.

The Autonomic Nervous System and Heart Rate

The opposite effects of the sympathetic and parasympathetic divisions are clearly illustrated in the case of the regulation of heart rate (see Figure 5-23). As you will learn in a later chapter, the rate at which the heart beats can be altered in order to meet varying demands for blood that are presented by the body's organs. Heart rate is increased by the sympathetic division and decreased by the parasympathetic division. Changes in heart rate are initiated by nerve centers in the brainstem that communicate with the sino-atrial node of

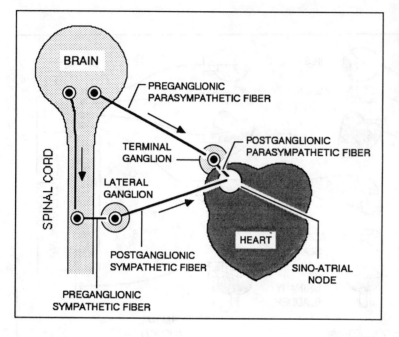

Figure 5-23
Sympathetic and parasympathetic pathways linking the brain with the sino-atrial node of the heart. See text for discussion.

the heart. Slowing the heart down is achieved by increasing the frequency with which impulses are sent via pre- and post-ganglionic fibers of the parasympathetic division, the fibers being constituents of the tenth cranial nerve (i.e., the Vagus nerve).

An increase in heart rate is effected by increasing the frequency of impulses reaching the sino-atrial node over sympathetic pre- and post-ganglionic fibers. As seen in Figure 5-23, this is achieved by first sending impulses down through the spinal cord to the level of the heart via association fibers that are components of a descending tract. At the level of the heart, the impulses are transmitted to the cell bodies of pre-ganglionic fibers in the grey matter. The circuit is then completed as the pre- and then post-ganglionic fibers relay the nerve impulses to the sino-atrial node.The sino-atrial node either increases or decreases the heart rate according to the relative balance of acetylcholine and nor-epinephrine released at the node by the post-ganglionic fibers.

The Adrenal Medulla

Among the tissues innervated by the sympathetic division, is the medulla (central portion) of the adrenal glands. Only preganglionic fibers innervate the adrenal medulla (i.e., there are no post-ganglionic fibers; see Figure 5-22). In response to these impulses, the adrenal glands release two closely-related hormones into the bloodstream: epinephrine and nor-epinephrine (i.e., the same substance that is released by the post-ganglionic fibers of the sympathetic division). These two hormones serve to augment the responses of organs also receiving sympathetic innervation, as well as influence tissues not receiving any sympathetic innervation at all. Moreover, the hormones circulate in the bloodstream for some period of time, so that adrenal secretion has a longer lasting effect than does the discharge of nerve impulses by the post-ganglionic fibers of the sympathetic division.

CHAPTER 6

THE RECEPTOR SYSTEM

Our perception of ourselves and the universe around us relies on the ability of the body's receptor system to transform chemical and physical stimuli into nerve impulses that are carried to the central nervous system for proper interpretation. The cells and tissues that comprise the receptor system may be divided into classes according to the nature of the stimulus that they detect. Accordingly, there are (1) **mechanoreceptors** (which respond to touch, pressure, and to vibration), (2) **thermoreceptors** (which respond to changes in temperature), (3) **photoreceptors** (which respond to light), (4) **chemoreceptors** (which respond to a variety of chemical stimuli), and (5) **nocioceptors** (which respond to physical or chemical damage).

Receptors may also be classified according to their location in the body. Using this criterion, three major classes may be identified: (1) **interoceptors**, (2) **proprioceptors**, and (3) **exteroceptors**. The interoceptors (also called **visceral receptors**) are responsible for sensing stimuli that arise deep within the body. For example, sensations of pressure arising from the organs of the digestive system or from the urinary bladder are the result of stimulation of interoceptors in these organs. On a more subtle scale, interoceptors in the walls of certain blood vessels serve to sense blood pressure or the oxygen and carbon dioxide contents of the blood. Sensations of hunger and thirst are also considered visceral senses.

Proprioceptors are located in the skeletal muscles and also in joints and are responsible for our awareness of muscle position and movement. Finally, the exteroceptors are responsible for our awareness of stimuli at the body surface or at some distance from the body. Included here are the receptors of the skin (i.e., the **cutaneous receptors**) and the so-called **special receptors** that are responsible for the senses of *vision* (i.e., the eyes), *hearing* (i.e., the ears), *gustation* or taste (i.e., the tongue), and *olfaction* or smell (i.e., the nose).

KINESTHESIA

If you were asked to close your eyes and slowly raise up your hand to touch your nose with your index finger, you probably could perform this task with relative ease. Your ability to do this rests with your constant awareness of the position of the different parts of your body (in this case your arm, hand, and nose) and their motion with respect to one another. This sense is known as **kinesthesia** and it relies on the properties and actions of the body's proprioceptors.

There are two major kinds of proprioceptors; these are the **Golgi tendon organs** and the **muscle spindle apparatuses**. Both are mechanoreceptors. The Golgi tendon organs are located in tendons (tendons join one muscle to another and also join muscle to bone). Each time that a muscle contracts, the tendon tissue at the ends of the muscle is stretched, and this stretching acts as a mechanical stimulus to the receptor cells. The stimulated receptors respond by triggering nerve impulses in the sensory fibers with which each Golgi tendon organ is associated. The long dendritic processes of these fibers are myelinated, so that the sensory impulses reach the central nervous system very quickly.

Each muscle spindle apparatus consists of several tapered receptor cells, called **intrafusal fibers**, enveloped in a thin layer of connective tissue. These assemblies are buried among the "normal" striated cells (or **extrafusal fibers**) near the ends of muscles, where the muscle fibers give way to the elastic tissue of the attached tendons. Within each spindle apparatus, the receptor cells are oriented with their long axis parallel to the axis of muscle contraction (Figure 6-1). The dendritic endings of sensory nerve fibers are twisted around each of the spindle cells to form a helical belt. The stretching of a muscle (for example, as you carry a briefcase or book at your side) stretches and thereby stimulates the intrafusal cells. The receptors then stimulate the spiral belt of dendritic endings of the associated sensory nerve fibers. The result is the flow of nerve impulses over the sensory fibers to the central nervous system, where the number and frequency of the incoming impulses is used as a measure of the amount of muscle stretching that is taking place. The stretching of the muscle is followed by reflexive contraction of the muscle in order to return the muscle to its normal length.

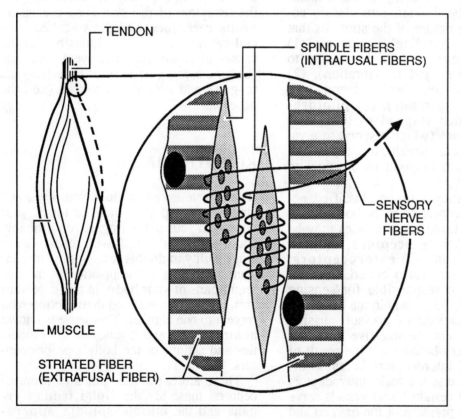

Figure 6-1
The **muscle spindle apparatus**. These receptors are located in the tendons that link muscle to bone; they consist of **spindle fibers** (**intrafusal fibers**) around which are twisted the dendritic endings of sensory neurons. When a muscle is stretched, the spindle fibers are mechanically stimulated; they then stimulate the associated sensory neurons. See the text for additional details.

Proprioceptors also play an important role in determining the extent of muscle contraction that is necessary in order to lift or move an object. For example, far less contraction of the biceps muscle is needed in order to lift a magazine off the table than is needed to lift a heavy book. As the muscle attempts to raise the object off the table, the resistance offered by the object's weight brings about an initial stretching of the muscle and this stretching is sensed by the proprioceptors. The greater is the weight (and resistance) of the object, the greater is the degree of initial stretching, and the greater is the number and frequency of sensory impulses arising in the sensory fibers associated with the proprioceptors. This sensory information is used as a gage to determine the intensity of the motor output that is necessary in order to bring about sufficient contraction in the muscle to lift the entire weight. Thus, the degree of contraction is no less or greater than that actually needed.

EXTEROCEPTION

Most thoroughly studied and best understood among the receptors of the body are those involved in exteroception, that is sensing stimuli that occur at the body surface or that originate at some distance away from the body. The exteroceptor senses are divided into two groups: the **cutaneous senses** and the **special senses**.

CUTANEOUS SENSES

There are three classes of cutaneous receptors. Those that give rise to sensations of touch and pressure are called **tactile receptors**; those that give rise to sensations of heat or cold are called **thermoreceptors**; and those that give rise to sensations of pain are called **pain receptors**.

Tactile Receptors

The skin is a complex tissue (Fig. 6-2) and contains a variety of different tactile receptors. Those receptors that respond to mild disturbances of the skin (such as that caused by the weight of an insect or a small piece of paper) include **free nerve endings**, **Meissner's corpuscles** (cell-encapsulated nerve endings), and **hair-end-organs** (Figs. 6-2 and 6-3). Hair-end-organs are the dendritic endings of sensory nerve fibers that are wound around the shaft of a hair, near the hair's root. Movement of the hair mechanically stimulates these nerve endings and gives rise to a sensation of touch. (Use the tip of a pencil to move a hair on the back of your hand and you will sense this mild disturbance.)

Pacinian corpuscles are also tactile receptors. However, pacinian corpuscles respond to more vigorous deformation of the skin, such as that which occurs when a heavier weight (e.g., a book) is placed on the skin. Pacinian corpuscles are therefore said to respond to *pressure* rather than to touch. Like Meissner's corpuscles, pacinian corpuscles are encapsulated nerve endings (Fig. 6-3); however, the encapsulation takes the form of several successive layers of connective tissue cells, so that the organization is much like the concentric arrangement of layers in an onion. As you might expect, tactile receptors are not distributed uniformly over the body surface. Instead, some regions of the skin (e.g., the fingertips and lips) contain a much higher concentration of receptors than do other regions. Consequently, these receptor-rich areas are much more sensitive to touch and pressure.

Thermoreceptors

There are two major types of thermoreceptors: those that respond to *decreasing* skin temperature (usually called cold receptors) and those that respond to *increasing* skin temperature (usually called heat receptors). Structurally, the two types of thermoreceptors are represented for the most part by the endings of the dendritic branches of sensory nerve fibers.

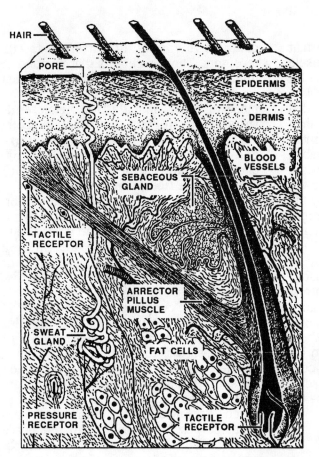

Figure 6-2
A section through the skin showing some of the receptors that respond to touch and pressure. Also shown are hairs that arise from the subepidermal layers and the sebaceous (i.e., oil) and sweat glands.

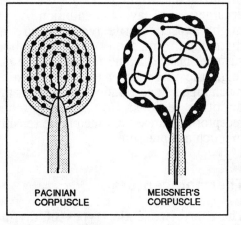

Figure 6-3
Pacinian and **Meissner's corpuscles**. These receptors (which respectively respond to pressure and light touch) consist of dendritic terminals of sensory neurons that are encapsulated in one or more layers of cells.

Pain Receptors

Most pain receptors are free dendritic nerve endings in the skin. These endings are stimulated by injury to the surrounding tissue. The release of the substances **bradykinin** and **histamine** by damaged cells has been linked to the stimulation of pain receptors. In this sense, the pain receptors are acting as chemoreceptors. Intense pressure applied to the skin and unusually high (or low) temperatures also give rise to sensations of pain and are believed to stimulate the free nerve endings directly.

Adaptation

You probably have had the experience of entering a swimming pool and finding the water uncomfortably cold at first; but after a few minutes, the water did not seem as cold. This phenomenon is known as **adaptation**. What happened was that the initial submersion of the skin in the pool water stimulated cold receptors and gave rise to sensations of cold. Although the pool temperature did not change, the skin's cold receptors ceased to be stimulated. As a result, sensory nerve impulses apprising you of the coldness of the water were halted. A similar effect is noticed when you place a watch on your wrist or a ring on your finger. At first, you are aware of the presence of the watch or ring, as tactile receptors in the skin are stimulated and trigger sensory nerve impulses. After a few minutes, they cease being stimulated and your awareness of the watch and ring subside. You may have heard stories about near-sighted people who raise their eye-glasses up to their forehead while they read a newspaper, and then later search everywhere for their glasses. (After several minutes they no longer feel the weight of their glasses on their forehead and are unaware of the presence of the object.)

The phenomenon of adaptation is important. Because adaptation does not begin until the stimulus has been applied continuously (and unchanged) for a short period of time, you are made aware of the stimulus' presence and have the option to react to it in some way. For example, sensing that a fly has settled onto your skin, you may elect to brush it away. Sensing that the pool water is quite cool, you may decide to get out of the pool. However, if the stimulus persists unchanged for some time, your awareness of it abates because the receptors adapt and fail to generate sensory signals. This frees the central nervous system to attend to the multitude of other sensory signals. Were

this not so, you would be preoccupied continuously by the weight and feel of your clothing, eye-glasses, rings, watches, shoes, and so on.

Different receptors adapt at different rates. As illustrated above, tactile and thermoreceptors adapt very quickly. On the other hand, pain receptors (and also proprioceptors) adapt very slowly. It is important to remember, that the adaptation that takes place involves the receptors–*not the central nervous system.*

THE SPECIAL SENSES

There are four special senses: **gustation**, the sense of taste; **olfaction**, the sense of smell; **vision**, the sense of sight; and **audition**, the sense of hearing.

GUSTATION

Gustation (the sense of *taste*), and olfaction (the sense of *smell*.) are closely related to one another. In fact, it is often difficult to distinguish between taste and smell. For example, when eating a meal, the "flavor" that is sensed is the result of stimulation of *both* the gustatory receptors and the olfactory receptors. Moreover, you have probably noticed that when you have a head cold and the nasal passageways are blocked, food seems to lose much of its taste.

The gustatory receptors are examples of chemoreceptor. They are stimulated by chemical substances that dissolve in the saliva. These receptors are clustered together to form a number of small **taste buds** scattered primarily over the surface of the tongue but also present on the palate and the walls of the pharynx (Fig. 6-4). Each taste bud contains about twenty receptor cells and opens onto the surface of the tongue through a small pore. Hair-like projections from the uppermost receptor cells project through the pore into the oral cavity. Each of the receptor cells is innervated by the dendritic endings of sen-

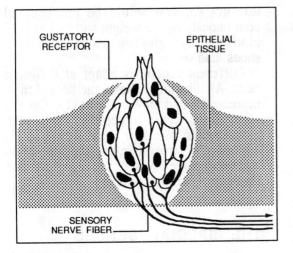

Figure 6-4
Organization of gustatory receptors to form a taste bud.

sory nerve fibers. These fibers conduct impulses to the brainstem through the *facialis* and *glossopharyngeal* nerves (cranial nerves VII and IX; see Table 5-1). From the brainstem, another family of fibers carries the sensory information to the cerebral cortex; other fibers carry the signals to the medulla, where autonomic pathways stimulate the salivary glands and augment the secretion of saliva.

It is generally agreed that there are four different "taste qualities:" *sweet*, *sour*, *salty*, and *bitter*. Chemicals that taste sour are acids, the degree of sourness varying in relation to the hydrogen ion concentration or pH that is created in the saliva by the substance. As you might expect, chemicals that taste salty are usually salts, such as sodium chloride. In addition to sugars, a variety of other chemical substances give rise to sweet sensations. Bitter tastes are also produced by a variety of chemical substances. Many poisonous substances taste bitter, a fact that has obvious survival value.

Not all regions of the tongue are equally sensitive to the four taste qualities. Rather, different regions respond to different substances; the regions of the tongue that are most sensitive to each of the four tastes are shown in Figure 6-5. Not all individuals

show equal sensitivity to specific substances, and this appears to be genetically determined. Individual variations are best illustrated in the case of PTC (phenylthiocarbamide). About 80% of the population finds that PTC tastes sour, whereas the other 20% finds that PTC has no taste at all. Although individual gustatory receptors are most sensitive to substances that produce a specific taste, receptors can be stimulated by more than one type of taste.

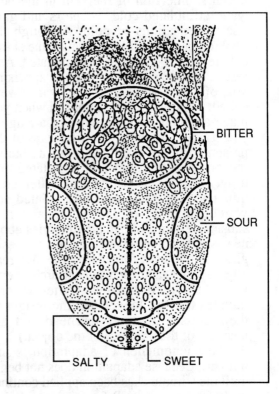

Figure 6-5
The surface of the tongue is not equally sensitive to all four taste qualities. Shown here are the areas most sensitive to bitter, salty, sour, and sweet tastes.

OLFACTION

Like the gustatory receptors, the olfactory receptors are chemoreceptors. The substances detected by the olfactory receptors dissolve first in the air that is inspired through the nose and then dissolve in the thin layer of mucus that covers the olfactory receptors.

The receptor cells form a small patch of olfactory membrane in a narrow cleft at the top of the nasal cavity (Fig. 6-6). During quiet breathing, the inhaled air does not flow over these receptors, although they can be reached by diffusion. More forceful inspiration or sniffing does bring the air into contact with the receptors and is followed by a strong olfactory sensation. The olfactory sense is more complex than the gustatory sense, and most individuals can distinguish thousands of different odors.

As seen in Figure 6-6, the receptors are elongate cells with small projections (called *cilia*) extending from the basal ends of the cells into the nasal cavity. At their apical ends, the receptors synapse with sensory fibers of the *olfactory* nerve (cranial nerve number I). Impulses are conducted by the olfactory nerve to the olfactory region of the brain (which is located on the underside of the cerebral cortex). Like many of the cutaneous receptors, the olfactory receptors exhibit adaptation. As a result, an odor is noticed for only a short period of time.

Figure 6-6
The olfactory membrane.   In a narrow cleft at the top of the nasal cavity (seen in the insert) is the olfactory membrane—the receptor apparatus that detects odors. Forceful inspiration or sniffing act to draw air onto the exposed endings of the olfactory receptor cells. Chemicals that are dissolved in the inspired air stimulate these cells, thereby giving rise to sensations of smell .

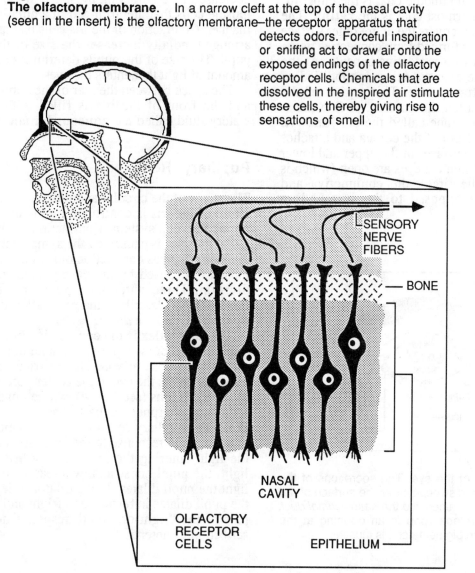

SENSORY NERVE FIBERS

BONE

NASAL CAVITY

OLFACTORY RECEPTOR CELLS

EPITHELIUM

VISION

The eyes are responsible for the sense of vision and consist of an elaborate apparatus that ultimately converts the energy of light rays into nerve impulses. To understand how the eyes function, let's begin by considering the organ's anatomy, noting the individual functions of the eye's various components.

Figure 6-7 is an exterior view of the eye as seen from the front, and Figure 6-8 shows the internal parts as seen in a side-view. The eye is a hollow, ball-shaped structure, the surface of which is formed by three coats of tissue. The outermost coat is the **sclera**; the middle coat is the **choroid**; and the innermost coat is the **retina**. The sclera, which covers the entire outer surface of the eye, is formed from connective tissue and serves as a protective capsule. At the front of the eye, the sclera is modified so that it is transparent. This region of the sclera is called the **cornea**. A fine transparent membrane called the **conjunctiva** covers the front of the cornea and attaches to the undersurfaces of the upper and lower eye-lids. When the eyes are open (which is most of the time), the conjunctiva and cornea are exposed to the surrounding air. To prevent these delicate tissues from becoming dry, their surfaces are kept moist by the flow of tears from the **lacrimal glands** (Fig. 6-7). The tears flow across the front of the eyes and are drained by the **nasolacrimal ducts**, which empty the fluid into the nasal cavity.

When looking at the front of the eye, the donut-shaped iris muscle can be seen behind the cornea. This muscle, which is rich in pigment and gives rise to the characteristic eye colors (e.g., brown eyes, blue eyes, green eyes, etc.), is a part of the middle coat of the eye–the **choroid** coat. The hole at the center of the iris is called the **pupil**; the size of this opening can be decreased by contracting those iris smooth muscle cells that are arranged in a circular manner. Contraction of the iris cells that are arranged radially increases the size of the pupil. The size of the pupils determines the amount of light that enters the eyes.

The space between the rear of the cornea and the front of the iris is filled with a watery fluid called the **aqueous humor**.

Pupillary Reflexes

The status of the circular and radial musculature of the iris is controlled by the autonomic nervous system. Contraction of the radial muscles is caused by the sympathetic division, whereas contraction of the circular muscles is caused by the parasympathetic division. Three autonomic pupillary reflexes may be described: the *near reflex*, the *light reflex*, and the *dilator reflex*.

The near reflex is the constriction of the pupil that occurs when looking at an object that is very close to you. This restricts the light that enters the eye to the center portion of the eye's **lens** (see below), where image focus and sharpness is best. The light reflex is the pupillary constriction and dilation that occurs in response to the brightness of the light entering the eye. In very bright light, the pupil constricts, whereas is dim light the pupil dilates. In the dilator reflex, the pupil dilates in response to fright and is part of the "fight-or-flight" response described in Chapter 5.

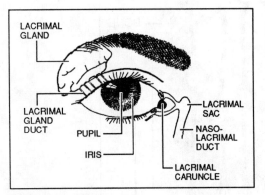

Figure 6-7
Front view of the eye. The secretions of the lacrimal gland sweep across the surface of the eye and are drained into the *nasolacrimal duct*. The pupil (black spot) is an opening at the center of the pigment-rich iris muscle.

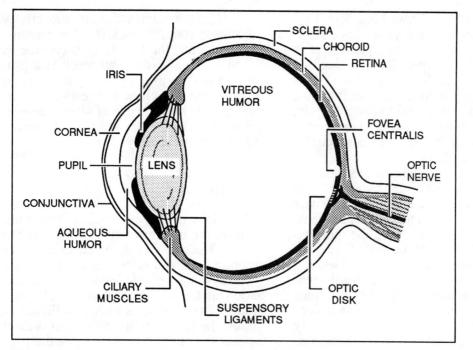

Figure 6-8
A side view showing the internal organization of the eye. See the text for details.

The Lens and Accommodation

Behind the iris, the choroid coat forms the **ciliary muscles** (Fig. 6-8). Tiny ligaments (called **suspensory ligaments**) arising from the ciliary muscles suspend the eye's **lens** in position behind the pupil. The lens is a flexible structure whose natural shape is round, much like a ball. However, the tension in the suspensory ligaments is high and the resulting pull on the lens' circumferential edges acts to flatten the lens, altering its shape to that of a gently biconvex disk. Contraction of the ciliary muscles pulls the choroid coat forward a tiny distance, and this serves to introduce slack into the suspensory ligaments. Under these conditions, the natural elasticity of the lens causes it to become rounder. This relationship between the shape of the lens and the status of the ciliary muscles is fundamental to the mechanics of accommodation; therefore, the essential points bear repeating:

(1) by pulling the choroid coat forward, contraction of the ciliary muscles loosens the suspensory ligaments, and allows the lens to become rounder;
(2) relaxation of the ciliary muscles allows the choroid coat to slide backwards, increasing the pull on the suspensory ligaments and flattening the lens.

Lens shape and distance of focus. A simple experiment that you can quickly perform illustrates the relationship between lens shape, ciliary muscle action, and the focusing of light rays entering the eye. Look out of the window or across the room at an object that is 20 of more feet away. Now close your eyes for several seconds and re-open them. You should note that immediately upon opening your eyes, the object is in focus. Now, hold this page about eight inches from your eyes and focus on the print. Close your eyes, and after several seconds re-open them. Notice that this time the object (i.e., page of print) was *not* immediately in focus and that it takes a small fraction of a second to bring the print into focus. You might also be able to sense the muscle strain that accompanies focusing on such a close object.

What do these observations demonstrate? Focusing on a distant object requires a flattened lens, whereas focusing on a near object requires a round (biconvex) lens. Indeed, the nearer the object, the more biconvex the lens must be (see Fig. 6-9). Focusing on a distant object occurs without ciliary muscle contraction because the tension in the suspensory ligaments flattens the lens. In contrast, focusing on a near object requires contraction of the ciliary muscles; their contraction slackens the suspensory ligaments, thereby allowing the lens to assume a biconvex shape. The process by which the shape of the lens is changed in order to keep an image properly focused on the photoreceptor layer at the rear of the eye is called **accommodation**.

As a person ages, the lenses of his eyes progressively lose their elasticity. As a result, although contraction of the ciliary muscles introduces slack into the suspensory ligaments, the lens fails to become as round as it once did. The result is a progressively deteriorating ability to focus on objects that are close to the eyes. Thus, most people over the age of 40 require eyeglasses which place an additional biconvex lens in front of each eye.

The Retina

The lens of the eye focuses incoming light on the innermost coat of the eye–the **retina**. The retina contains the complex of receptor cells and nerve cells that convert light rays into nerve impulses. The organization of the several layers of cells that comprise the retina is shown in Figure 6-10. At the surface of the retina that is flush against the choroid coat, there is the **pigment layer**. This is a layer of cells rich in the dark pigment **melanin**. The role of the pigment layer is similar to that played by the black coat of paint that covers the inside surfaces of a camera–namely, to prevent the reflection of light rays. Thus, any light not absorbed by the photoreceptor cells themselves is absorbed by the pigment layer.

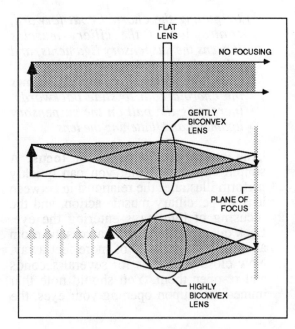

Figure 6-9
Effects of curvature on the focusing power of a lens. In this figure, light reflected from the surface of an object (i.e., an arrow) is directed into each of three lenses. Whereas a flat lens has no focusing power at all (*top*), a biconvex lens focuses the light reflected from the arrow on a plane located some distance behind the lens (*middle*). The more biconvex the lens (*bottom*), the closer the object may be to the lens while still remaining in focus. Notice that the image of the object created on the focal plane is *inverted*.

Seated on the pigment layer is the layer of photoreceptor cells. Two classes of photoreceptors are present; they are called **rods** and **cones** (the names describe the general shapes of the cells, the the rods being rod-like and cones being conical). When light enters the rods and cones, it is absorbed by special pigment molecules inside of these cells. Light absorbtion by the pigments triggers a series of metabolic reactions that eventually results in the stimulation of the dendritic endings of nerve cells associated with the photoreceptors.

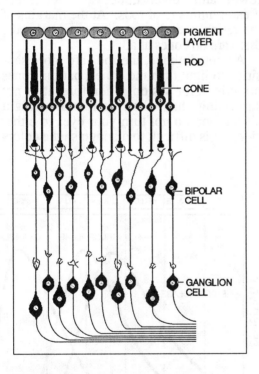

Figure 6-10
The retina. The pigment layer is appressed against the choroid coat. Seated on the pigment layer is the layer of rods and cones (i.e., the photoreceptors). The photoreceptors stimulate bipolar nerve cells and these transmit impulses to ganglion cells. The long axonal processes of the ganglion cells converge on the optic disk and enter the optic nerve. It should be noted that light must pass through the ganglion cells and the bipolar cells *before* reaching the photoreceptor layer.

The rods and cones have different properties and function differently in vision. The rods are the more sensitive photoreceptors and function in dim light, although they do not provide much image detail or information about color. In contrast, the cones function in bright light and provide detailed images that possess color.

The light-absorbing substance of rods is a purple material called **rhodopsin** (or **visual purple**). It is made up of the pigment, **retinene**, and the protein, **opsin**. When rhodopsin absorbs light, it breaks down into its component parts, retinene and opsin. This chemical reaction leads to a change in the polarity of plasma membranes of the rods (and the stimulation of associated nerve endings). Some of the retinene recombines with opsin, thereby regenerating the rhodopsin. The remainder of the retinene is reduced to **vitamin A**, some of which enters the bloodstream and is degraded. Retinene lost in this way must be replaced, and under normal circumstances is resynthesized from vitamin A in the diet.

Bright light causes the rapid breakdown of rhodopsin and temporarily renders the rods functionless. This explains why a person entering a darkened theater from daylight finds it extremely difficult to see. Gradually, the rhodopsin is replenished and vision returns. This phenomenon is called "dark adaptation."

When reentering daylight from a darkened theater, the reverse effect is noted. Initially, the daylight seems painfully bright, but as the rhodopsin produced during dark adaptation is quickly degraded, normal daylight vision returns. As you might expect, this phenomenon is called "light adaptation."

Although the rods provide sensitivity to low levels of light, they do not provide any information about color. Color vision is a function of the cones. Although most individuals can distinguish hundreds of different colors and shades, there are only three different kinds of cones: red-sensitive cones, green-sensitive cones, and blue-sensitive cones. Each type of cone possesses a different kind of light-absorbing pigment.

The red-sensitive family of cones contains the pigment **erythrolabe**; the green-sensitive family of cones contains the pigment **chlorolabe**; and the blue-sensitive family of cones contains the pigment **cyanolabe**. Our ability to distinguish red, green, and blue objects depends upon which family of cones is stimulated by the light reflected by the object. For example, light that is reflected into the eyes from a red object is differentially absorbed by the pigment molecules of the red-sensitive cones; light that is reflected into the eyes from a green object is differentially absorbed by the pigment molecules of the green-sensitive cones, and so on. As shown in Figure 6-11, there is some overlap in the color sensitivities of the three cone families. Absorbtion of light by the pigment molecules sets off the metabolic reactions that culminate in the stimulation of nerve fibers associated with the cone cells. Our perception of colors other than red, green, and blue stems from the stimulation of combinations of red-, green- and blue-sensitive cones. For example, when yellow light falls on the retina, both red- and green-sensitive cones are stimulated. The blue-sensitive cones are not stimulated at all by the yellow light. If light falling on the retina stimulates all three types of cones equally, the color is interpreted as being white.

Color-Blindness. Color-blindness is an abnormality in which a person either fails to see color altogether or has difficulty distinguishing certain colors. "Complete color-blindness" is very rare, whereas "red-green color-blindness" is quite common (about eight percent of males, but less than one percent of females are red-green color-blind). The abnormality (which is inherited via the sex chromosomes) occurs when one of the three families of cones is either absent or poorly functioning. Usually, the defect involves either the red-sensitive cones or the green-sensitive cones. In red-green color-blindness due to malfunctioning (or absent) red-sensitive cones, red light falling on the retina is absorbed by and weakly stimulates the green-sensitive cones. For such a person, a red object is difficult to distinguish from a green object.

Distribution of Rods and Cones in the Retina. The photoreceptor layer of the retina consists of millions of cones and rods; however, these cells are not uniformly distributed in the retina. Instead, the cones are concentrated in a small area at the center of the retina called the **fovea centralis**. As seen in Figure 6-8, the fovea centralis forms a small depression in the retina and lies directly in line with the center of the cornea and the center of the lens. As one moves away from the fovea, one finds fewer and fewer cones and relatively greater numbers of rods. At the margins of the retina, the photoreceptors are all rods; there are no cones at all.

As noted above, the rods are more sensitive to light than the cones, but the cones provide color vision. Therefore, it is not surprising that when you look straight ahead (and do not turn your eyes to either side), it is difficult to distinguish the colors

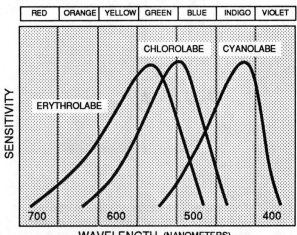

Figure 6-11.
Action spectra of the three types of pigments found in the three families of cones. The red-sensitive cones contain the pigment erythrolabe; the green-sensitive cones contain chlorolabe; and the blue-sensitive cones contain cyanolabe. Note that there is some overlap among the spectra.

of the objects at the edges of your field of view. By the same token, in very dim light, one can more easily detect objects at the edges of the field of view than at the center.

Nerve Cells of the Retina. The cones and rods form junctions with a layer of nerve cells called **bipolar cells** (see Fig. 6-10). When the photoreceptors are stimulated by light, they cause nerve impulses to be propagated in these bipolar cells. Whereas each cone forms a junction with a different bipolar cell, two or more rods may form a junction with a single bipolar cell. Because there is a 1:1 relationship between cones and bipolar cells, the cones provide greater visual *acuity*. The bipolar cells are quite short, and within the retina, they transmit their nerve impulses to a second layer of nerve cells called **ganglion cells**. The long axons of the ganglion cells run over the surface of the retina and converge upon a particular point on the retina known as the **optic disk**. At the optic disk the axons are bundled together to form the **optic nerve**, which exits at the rear of the eye. The optic disk is also the entry point for blood vessels that nourish the eye.

Because the optic disk contains no cones or rods, any light that falls on this region of the retina goes undetected. Consequently, the optic disk is also referred to as the eye's **blind spot**. You can readily demonstrate the blind spot to yourself in the following way. Draw a dark "x" in the corner of a small sheet of white paper, and while keeping one eye closed, hold the paper several inches in front and slightly to the side of your other (open) eye. Look straight ahead (not at the "x") and slowly move the paper in a circle. You should find that there is a position in which you can no longer detect the "x," even though you can still see the surrounding white paper. In that position, the image of the "x" is falling on the blind spot. Normally, a person is unaware of the existence of the blind spot because both eyes are open and it is not possible for an image to fall on the blind spots of both eyes *at the same time*.

The surface of the retina is bathed by a viscous fluid called the **vitreous humor**, which fills the space between the rear of the lens and the retina's surface (Fig. 6-8).

Visual Areas of the Brain

Figure 6-12 shows the nerve pathways that lead from the retinas of the two eyes to the visual areas of the brain. As seen in the figure, the fibers that originate in the nasal half of each of the two retinas cross over in the **optic chiasma** to the opposite side of the brain. In contrast, fibers that originate in the marginal (i.e., outer) half of each retina enter the optic chiasma but do not cross over to the opposite hemisphere. The fibers that exit at the rear of the optic chiasma create the left and right **optic tracts**. Each optic tract contains fibers that originate in the retinas of *both* eyes. For example, the left optic tract contains fibers that originate in the outer half of the left eye and the nasal half of the right eye.

The crossing over of certain fibers in the optic chiasma has several implications. Clearly, if the left optic nerve is destroyed by injury or disease, vision in the left eye is totally lost. However, destruction of the left optic tract leaves a person with partial vision in both eye. That is, the affected person would retain vision in the nasal half of the left eye and the outer half of the right eye. (Similarly, if the right optic tract were destroyed, vision would be retained in the nasal half of the right eye and the outer half of the left eye.)

Common Eye Abnormalities

In a person with normal vision, the eye lens is smooth and clear, and the changes in lens shape that occur during accommodation ensure that the entire image that is being viewed is properly focused onto the retina. There are, however, a number of common eye abnormalities in which the image is not properly focused on the retina; among these are the conditions known as **far-sightedness** (also known as **hyperopia**), **near-sightedness** (also known as **myopia**), and **astigmatism**.

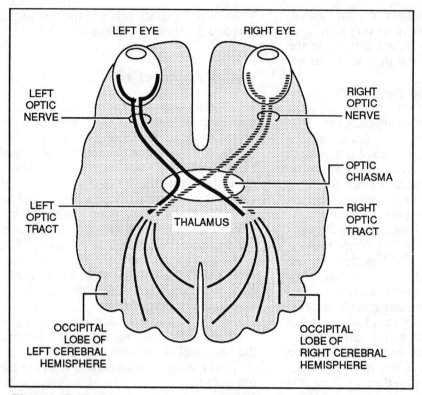

LEFT EYE   RIGHT EYE

LEFT
OPTIC
NERVE

RIGHT
OPTIC
NERVE

OPTIC
CHIASMA

LEFT
OPTIC
TRACT

THALAMUS

RIGHT
OPTIC
TRACT

OCCIPITAL
LOBE OF
LEFT CEREBRAL
HEMISPHERE

OCCIPITAL
LOBE OF
RIGHT CEREBRAL
HEMISPHERE

Figure 6-12
Nerve pathways that lead from the retinas of the two eyes to the brain. The fibers that originate in the nasal half of each retina cross over in the **optic chiasma** to the opposite side of the brain. In contrast, fibers that originate in the outer half of each retina remain on the same side of the brain. Therefore, each side of the brain receives visual signals from both eyes. See the text for additional details.

Far-sightedness. The correct focusing of light onto the retina is illustrated in the *top* section of Figure 6-13. However, in many individuals (especially people who are over 40 years old) who view an object that is only a short distance from the eyes, the focusing power of the lens is too low to focus the image on the retina. Instead, the image created on the retina is "not yet" in focus (i.e., it is focused on a hypothetical plane that lies a short distance *behind* the retina; *middle* section of Fig. 6-13). This condition usually results from one of the following two problems; either (1) the eye is too short (i.e., the distance between the rear of the lens and the surface of the retina is too short), or (2) the lens has lost some

of its elasticity and fails to become sufficiently round when the ciliary muscles contract (see earlier). The latter is usually the case in people who are over 40.

The condition just described is called **far-sightedness** (also **hyperopia**) because such individuals easily focus distant images. A distant scene does not require that the lens become so round in order for the image to be properly focused on the retina. Far-sightedness can be corrected with eye-glasses whose lenses are *convex* (Fig. 6-13).

Near-sightedness. A person who is **near-sighted** (or **myopic**) does not see distant images very clearly. Usually, this is

90

because the eye is too long (i.e., the distance between the rear of the lens and the surface of the retina is too great), with the result that the image is focused on a plane that lies *in front of* the retina (*bottom* section of Fig. 6-13). Near-sightedness can be corrected with eye-glasses whose lenses are *concave*.

Astigmatism. A common vision abnormality is **astigmatism**. In astigmatism, certain features of the image are in focus, while other features are not. For example, vertical features may be focused on the retina, whereas horizontal features may be focused in front of or behind the retina. The reverse may also occur (i.e., horizontal features in focus but vertical features out of focus). The most common sources of this abnormality are defects in the curvature of the cornea and irregularities in the smooth-

ness or clarity of the lens. Like hyperopia and myopia, many forms of astigmatism can be corrected using eye-glasses having appropriately shaped lenses.

HEARING (AND EQUILIBRIUM)

The ears are the source of several special senses. While the most obvious of these is the sense of hearing, the ear also provides the brain with information about head position and head movement. There are three major parts to the ear: the **outer ear**, **middle ear**, and the **inner ear**. Of these, it is the inner ear that contains the receptor apparatus for hearing (called the **cochlea**) and for head position and movement (called the **vestibular apparatus**).

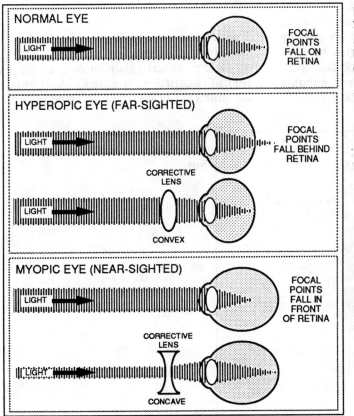

Figure 6-13
The most common eye abnormalities are **farsightedness** (i.e., **hyperopia**) and **near-sightedness (myopia)**. In the normal (*emmetropic*) eye, light is focused on the retina; however, in an unaided hyperopic eye, the plane of focus falls behind the retina, and the image seen is blurred. This condition may be corrected by placing a biconvex lens in front of the eye. The biconvex lens performs some preliminary focusing before the light enters the eye. In myopia, the focal plane falls in front of the retina. Again, the image seen is blurred. Myopia can be corrected by placing a biconcave lens in front of the eye. The biconcave lens diverges the light rays, thereby driving the eye's focal plane further to the rear.

Sound

To properly understand and appreciate the sense of hearing, it is necessary to understand what *sound* is. Sound is produced by the vibration of an object (e.g., a guitar string, a metal tuning fork, or your own vocal cords). As the object vibrates back and forth, it pushes against and then pulls back the adjacent layers of air molecules. The motion of the air is then transferred to neighboring molecules which repeat the cycle, thereby causing a wave that moves in all directions away from the original source of vibration. (A good analogy are the ripples produced when a small stone is dropped into a very still pond.) In the parlance of physics, a pressure wave is created radiating in all directions from the sound source. In the case of sound travelling through air, the waves typically travel at speeds greater than 1,000 feet per second. Thus, a sound made 2,000 feet away takes about 2 seconds to reach your ears. When travelling through materials other than air, the speed of the sound is altered.

Every sound wave has certain physical characteristics. Among these is **frequency** (or **pitch**), which is a measure of how rapidly the air is vibrating back and forth and is measured in **cycles per second** (each cycle is a complete back and forth movement) or **herz**. The higher the frequency, the higher the pitch. Humans are capable of detecting sound waves having frequencies as low as about 20 cycles per second and as high as about 20,000 cycles per second. Within this range, the ears can differentiate about 2,000 different pitches.

A second property of a sound wave is its **amplitude** (or **intensity** or **loudness**), which is measured in **decibels**. For example, a ticking watch has an intensity of about 20 decibels, ordinary conversation a value of about 60 decibels, and the sound of a car horn only 15 feet away about 100 decibels (the decibel scale is logarithmic). Sounds louder than about 150 decibels can be harmful to the ears. The subjective sensation of the loudness of a sound depends on the frequency of the sound as well as its amplitude, the ear being more sensitive to some frequencies than to others.

The Outer Ear

The outer (or external) ear (Fig. 6-14) consists of the **pinna** (or **auricle**) and the **auditory canal** (also known as the **external auditory meatus**). In humans, the pinna is not particularly large or highly

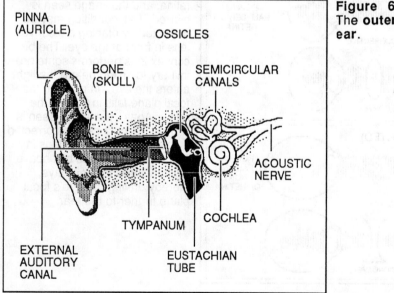

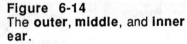
Figure 6-14
The **outer, middle,** and **inner ear.**

developed, whereas in many other mammals it serves as a horn to trap sound waves and can even be directed toward the source of the sound. In humans, the pinna is a flap of cartilage and skin that is usually flush (or nearly flush) against the side of the head; therefore, its ability to trap sound is considerably diminished. Sound passing into the pinna travels along the auditory canal (which passes slightly upward and backward for a distance of about one inch) and reaches the **tympanic membrane** (or **ear drum**). The tympanic membrane seals off the end of the auditory canal and serves to separate the outer ear from the middle ear. Sound waves striking the tympanic membrane cause the membrane to vibrate with the same frequency as the impinging sound. By its vibration, the tympanic membrane transmits the sound to the middle ear.

The auditory canal and the tympanic membrane are lined by skin. The skin of the auditory canal is rich in glands that secrete wax (called **cerumen**) onto the canal's surface. The wax lubricates the canal and rarely accumulates in amounts sufficient to interfere with normal hearing. Thus, one should not attempt to remove the wax by inserting objects into the canal and running the risk of damaging the tympanic membrane.

The Middle Ear

The middle ear is a small chamber in the skull (Fig. 6-15) containing three tiny bones called **ossicles**. The chamber is not completely closed, since the floor of the chamber contains a narrow opening that leads via the **Eustachian tube** into the rear of the mouth (i.e., the **pharynx**). Air can enter or leave the middle ear through this channel, and this permits equalibration of the gas pressure of the middle ear with the atmospheric pressure around us. Equalibration of pressure prevents rupturing the tympanic membrane when there is a change in external pressure (as when travelling by plane at great heights).

Normally, the Eustachian tube is closed; however, swallowing and chewing bring about a temporary opening of the channel, thereby allowing pressure equilibration. The Eustachian tube can be an avenue for passage of bacteria from the mouth into the middle ear, causing middle ear infections.

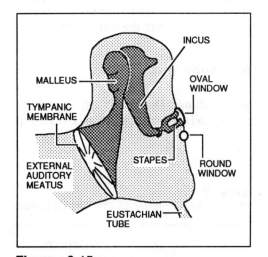

Figure 6-15
The components of the middle ear (see text for details). The narrow Eustachian tube leads downward from the air-filled middle ear into the rear of the mouth.

The far wall of the middle ear also contains two openings; these openings, however, are covered by thin membranes. The upper opening is the **oval window**, and the lower the **round window**. The ossicles are arranged to form a series of levers that transmit the vibrations of the tympanic membrane across the middle ear to the oval window. As we will see, the inner ear contains fluid that is set in motion by sound. Much more force is required to set liquid in motion than to set air in motion. The lever-like arrangement of the ossicles provides the means for the 20-fold increase in force needed to cause vibration of fluid in the inner ear (i.e., the ossicles act in a manner similar to a set of gears). Excessive movements of the ossicles are prevented by small muscles attached to the ossicles. The ossicle attached to the tympanic membrane is the **malleus**; the malleus articulates with the second ossicle, called the **incus**; finally, the incus moves the **stapes** back and forth, pushing in and out of the oval window. The oval window leads into a long, fluid-filled, twisted channel that eventually leads back to the round window. Thus, as the oval window is pushed inward by movement of the stapes, the round window bulges outward into the middle ear.

The Inner Ear

The Cochlea. The inner ear consists of the **cochlea, saccule, utricle**, and semi-circular canals (Fig. 6-16); these structures occupy a complex labyrinth in the temporal bone of the skull. The cochlea is concerned with the sense of hearing, whereas the other components of the inner ear allow us to sense head position and head movement; the latter senses are known as the **vestibular senses**. The cochlea is spiral-shaped chamber similar in appearance to a snail shell. Within the chamber are three fluid-filled channels called the **scala vestibuli**, the **scala media**, and the **scala tympani**. In Figure 6-17 the three channels are shown in an untwisted arrangement, so that the relationships among them and their relationship to the oval and round windows are more easily understood.

Figure 6-16
The inner ear consists of the **cochlea, saccule, utricle**, and **semi-circular canals**.

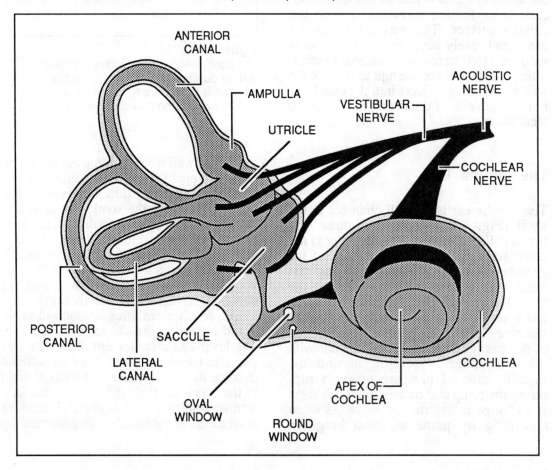

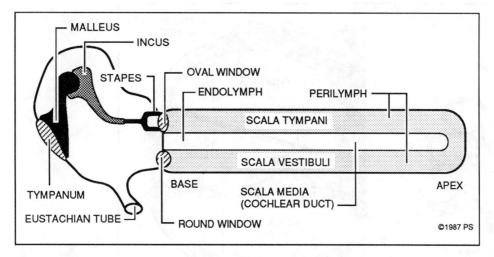

Figure 6-17
The three channels of the cochlea. The channels have been unwound, so
that their internal organization can be more clearly understood.

The scala vestibuli begins at the oval window near the *base* of the cochlea and winds its way up to the cochlea's *apex*, where it leads into the scala vestibuli; the scala vestibuli then winds its way back down to the cochlea's base, ending at the round window. The scala vestibuli and scala tympani make about two and one-half turns as they wind their ways between the apex and base of the cochlea. The fluid that fills these channels is called **perilymph**. Vibrations of the stapes are transmitted across the oval window so that the perilymph is set into motion.

The third fluid-filled channel of the cochlea, the scala media, is sandwiched between the scala vestibuli and scala tympani and is filled with a fluid called **endolymph** (Fig. 6-17). The floor of the scala vestibuli (which is also the roof of the scala media) is a thin membrane called the **vestibular membrane (also known as Reissner's membrane)**. A thicker structure called the **basilar membrane** serves as the floor of the scala media (which is also the roof of the scala tympani). The scala media houses the delicate machinery of sound detection called the **Organ of Corti**. Figure 6-18, which depicts a cross-section through the cochlea, shows the three channels as they wind their way upward from the base tot he apex.

The Organ of Corti. The organ of Corti (shown in detail in Fig. 6-19) contains the auditory receptor cells or **phonoreceptors**. Because hair-like projections emerge from the upper surfaces of these cells, the cells are also referred to as **hair cells**. The hair-like projections are embedded in the overhanging **tectorial membrane** which "floats" freely in the endolymph. The hair cells, which are estimated to be about 15,000 in number, are seated on the basilar membrane and form a succession of rows that begin at the base of the cochlea and extend up to the apex. In the basilar membrane, the hair cells form junctions with the dendritic endings of sensory nerve fibers. Collectively, these fibers form the **cochlear nerve**, which following its exit from the cochlea becomes a part of the **acoustic nerve** (i.e., cranial nerve number 8).

Movements of the oval window at the base of the cochlea cause the perilymph of the scala vestibuli to vibrate and this sets the vestibular membrane into vibration. These vibrations are transmitted across this thin membrane, so that the endolymph vibrates in sympathy. The vibrations of the endolymph then cause the basilar

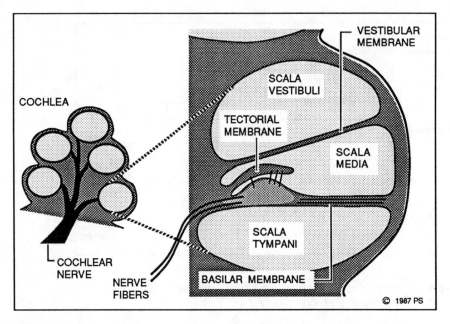

Figure 6-18
The two and one-half turns made by the cochlea as it winds its way upward from
the base to the apex are shown in the small insert. The exploded view is a section
across one of these turns, showing the scala vestibuli, scala media, and scala
tympani. The cochlear nerve runs up through the central axis of the cochlea,
giving rise to nerve fibers that form a spiral and radial (spoke-like) pattern.

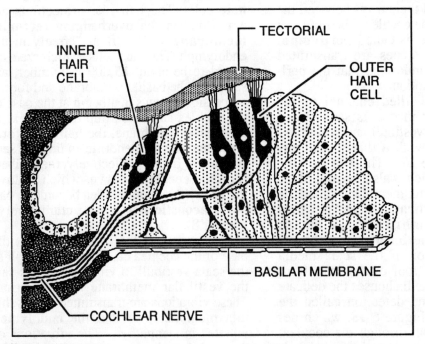

Figure 6-19
A magnified view of the **organ of Corti**. See the text for details.

membrane to vibrate. This membrane contains thousands of stiff fibers that radiate through the membrane from the cochlea's axis much like the spokes that radiate from the hub of a bicycle wheel to the wheel's rim. These stiff fibers are believed to have vibration properties similar to the reeds of a musical instrument. The lengths of the stiff fibers increase progressively as the basilar membrane winds its way up to the apex. Although the basilar membrane is capable of vibration over its entire length, it vibrates *maximally* at a specific point that is determined by the vibration frequency of the endolymph (and, therefore, the original source of the sound). The point of maximum vibration of the basilar membrane corresponds to the region of the membrane containing stiff fibers of the appropriate **resonant frequency**. Low frequency sounds vibrate the basilar membrane over its entire length, with the vibration peak occurring near the apical end of the scala media, where the stiff fibers are longer and have lower resonant frequencies. Movements of the basilar membrane caused by high frequency sounds peak closer to the basal end of the scala media where the stiff fibers are shorter and have higher resonant frequencies.

At the point of maximum movement of the basilar membrane, motion of the hair cells causes their stimulation, with the result that nerve impulses are generated in the sensory nerve fibers associated with these receptors. Impulses arising from a particular region of the Organ of Corti are conducted to a specific region of the auditory portion of the cerebral cortex and are interpreted as sound of a specific pitch. Most sounds are a complex mixture of pitches that stimulate several regions of the Organ of Corti. The resulting pattern of impulses received by the brain is sorted out, so that the interpretation matches the complexity of the original sound. The intensity or loudness of the original sound determines the amplitude of the vibrations of the tympanic membrane, the ossicles, the oval window, the cochlear fluids, and the basilar membrane. The greater the amplitude, the more

intense is the sensation. Some of the sensory neurons arising from the cochlea (and other regions of the inner ear) cross over in the brainstem. Therefore, both sides of the cerebral cortex receive input from each of the ears.

The Vestibular Apparatus

The **vestibular apparatus** consists of the **utricle, saccule,** and **semicircular canals** (Fig. 6-16) and is important in the maintenance of equilibrium. The apparatus provides information about head position and motion. Like the cochlea, the parts of the vestibular apparatus are filled with fluid.

The Saccule and Utricle. The saccule and utricle provide information about the head's position in space and are said to regulate *static equilibrium* and *linear acceleration*. Both structures contain sheets of hair cells, similar to those of the Organ of Corti (Figure 6-20). The projections (i.e., "hairs") ascend into a jelly-like layer containing small crystals of calcium carbonate called **otoliths**; the gelatinous layer is called the **otolith membrane**. At their basal ends, the hair cells associate with the dendritic endings of nerve fibers that make up the **vestibular nerve**. (The vestibular nerve and the cochlear nerve form the acoustic nerve; see above.)

When a person is standing upright, the hair cells of the utricle are oriented vertically, while the hair cells of the saccule are oriented horizontally. If that person suddenly moves forward (i.e., this would be *horizontal linear acceleration*), the inertia of the otoliths in the saccule causes them to lag behind, thereby pulling *backward* on the hair cell layer. This acts as a stimulus to hair cells in a specific region of the layer, causing them to generate nerve impulses in their associated sensory nerve fibers. These impulses are conducted to the brain where they are interpreted in the appropriate manner.

If a person standing upright on a chair suddenly steps down onto the floor (i.e.,

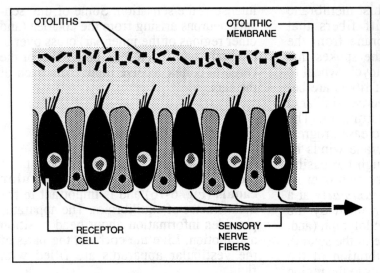

OTOLITHS

OTOLITHIC MEMBRANE

RECEPTOR CELL

SENSORY NERVE FIBERS

Figure 6-20
Organization of the
saccule (and **utricle**).
See text for explana-
tion.

this would be *vertical linear acceleration*), the inertia of the otoliths in the utricle causes them to lag behind, thereby pulling *upward* on the hair cell layer. As in the case of the saccule, this acts as a stimulus to hair cells in a specific region of the receptor layer, causing the cells to generate nerve impulses in their associated sensory nerve fibers. Upward and backward linear acceleration produce reciprocal effects to those of downward and forward linear acceleration.

Semi-Circular Canals. Arising from the end of the vestibular apparatus are the three fluid-filled (again, the fluid is endo-lymph) **semi-circular canals** (Fig. 6-21). These structures provide information about rotational movements of the head (or *dynamic acceleration*). Each canal is like a ring oriented in one of the three planes of space (i.e., "upward/downward" [the **anterior** or **superior canal**], "forward/backward" [the **posterior canal**], and "side-to-side" [the **lateral canal**]). Close to its point of origin, each canal possesses an expanded portion (called an **ampulla**) that houses the canal's receptor apparatus (known as a **crista**). The receptor consists of a layer of hair cells whose hair-like projections ascend into a gelatinous tuft called the **cupula**.

The cristae of the canals are stimulated

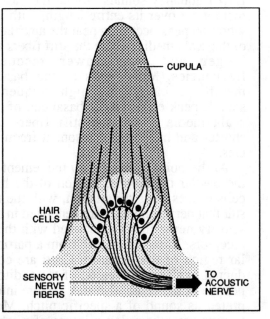

CUPULA

HAIR CELLS

SENSORY NERVE FIBERS

TO ACOUSTIC NERVE

Figure 6-21
Organization of an ampulla of a semi-circular canal.

by rotational acceleration within that canal's plane. For example, when the head is suddenly turned to the left in a horizontal plane, the fluid in the lateral canals lags slightly behind the canal's movement.

Relative to the moving canal, the fluid is rotating in the *opposite* direction. Rotation of the fluid pushes on the cupula bending it to one side and stimulating certain hair cells in the receptor layer. The receptor cells are associated with sensory nerve fibers, so that nerve impulses are conducted (via the vestibular and acoustic nerves) to the brain, apprising the brain of the head's motion. Upward (or downward) or sideways rotation of the head produce corresponding effects in the other two semi-circular canals.

THE CIRCULATORY SYSTEM

The circulatory system is the system of organs that is responsible for the circulation of the blood. Several major physiological roles are played by this system; these include

1. the transport of oxygen, nutrients, and raw materials to the body's tissues,
2. the removal of carbon dioxide and metabolic wastes from the body's tissues,
3. the transport of hormones and other substances from one tissue to another, and
4. the distribution of the body's heat of metabolism.

MAJOR COMPONENTS OF THE CIRCULATORY SYSTEM

The circulation of the blood requires a mechanism that keeps the blood in motion and an array of vessels through which the blood is conducted (Fig. 7-1). The motion of the blood is created by the pumping action of the **heart**. Blood is carried away from the heart by major vessels called **arteries**; indeed, an artery is defined as any major vessel that conducts blood away from the heart (note that the definition is unconcerned with the blood's composition, such as whether it is rich in oxygen or is oxygen-poor). It is the arteries that carry the blood into the body's major organs.

Once inside an organ, the arteries branch into a number of smaller vessels called **arterioles**, and the arterioles then branch into a **network** of finer vessels called **capillaries**. Whereas the walls of arteries and arterioles contain smooth muscle whose state of contraction or relaxation determines the blood pressure and regulates the flow of blood through the vessel, the capillary walls have no musculature. In fact, the walls of the capillaries are formed from a single, thin layer of cells (called **mesothelium**). It is the thinness and permeability of the capillary wall that permits the rapid exchange of materials between the blood passing through the capillaries and the surrounding tissue cells.

Blood is collected from the capillary networks by **venules**. A number of venules converge to form a **vein**, which then directs the blood out of the organ. The major veins return the blood to the heart. Like the arteries and arterioles, the venules and veins have muscular walls to regulate blood flow and blood pressure. The larger venules and the veins contain **valves**; these act like one-way gates that open only in the direction that leads to the heart. The action

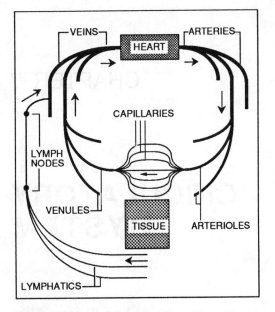

Figure 7-1
Major components of a circulatory system.
Such a system requires a pump (i.e., heart),
distributory vessels (i.e., arteries and arterioles), vessels for exchange (i.e., capillaries),
and collecting vessels (i.e., venules and
veins). In humans, an auxiliary system of collecting vessels exists (i.e., the lymphatics)
which empty their contents into major veins.

of the valves assures that the blood does
not back up in the circulatory system.

Also considered part of the circulatory
system are the **lymphatic vessels**. These
vessels arise *blindly* within a tissue (Fig. 7-
1) and act to conduct fluid toward the heart.
The fluid carried by the lymphatics is called
lymph (not blood). Like the veins and
larger venules, large lymphatics contain
valves. At intervals along the larger lymphatics are the **lymph nodes** which act to
filter the lymph (see later).

THE PULMONARY AND SYSTEMIC CIRCULATIONS

The human body contains two separate and
distinct circulations of the type described in
Figure 7-1. The two circulations are called

the **pulmonary circulation** and the
systemic circulation (Fig. 7-2). Each
circulation has its own pump (i.e., heart)
and its own set of arteries, arterioles,
capillaries, venules, and veins. Only one
organ of the body is served by the
pulmonary circulation and that is the *lungs*.
The sole purpose of this circulation is to
enrich the blood's oxygen content and to
diminish its content of carbon dioxide. This
testifies to the critical role of oxygen in the
body's metabolism and the continuous need
to eliminate the waste carbon dioxide that is
produced during metabolism.

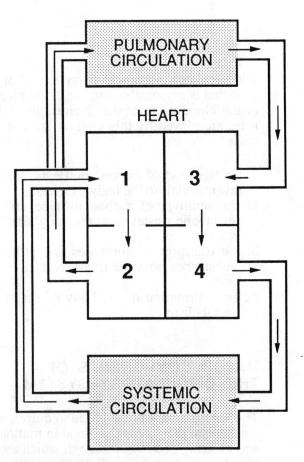

Figure 7-2
There are two blood circulations in the body.
The **pulmonary circulation** is fed by the
right side of the heart and directs blood
through the lungs only. The **systemic
circulation** is fed by the left side of the heart
and directs blood through all body tissues.

102

The systemic circulation carries blood to *all* of the body's organs (muscles, brain, liver, stomach, kidneys, etc.). This implies that the lungs are *also* served by the systemic circulation; that is, there are *two* circulations of blood in the lungs. The pulmonary circulation carries blood through the lungs in order to oxygenate the blood and eliminate carbon dioxide, while the systemic circulation carries blood through the lungs in order to nourish the lungs and carry away the lungs' wastes. The two bloods do not mix in the lungs.

CIRCULATION OF BLOOD THROUGH THE HEART

The pumps that pump blood through the pulmonary and systemic circulations comprise the right and left sides of the heart. We'll begin our consideration of the circulatory system by examining the organization and functions of the different parts of this most important organ.

The circulation of blood through the heart is shown in Figure 7-3. Blood returning to the heart from the *systemic* circulation enters the *right* side of the heart through two major veins: the **superior vena cava** (structure number *1* in Fig. 7-3) and the **inferior vena cava** (structure number *2* in Fig. 7-3). The superior vena cava returns blood to the heart from the upper parts of the body (head, neck, arms, upper chest, etc.), whereas the inferior vena cava returns blood to the heart from the lower parts of the body (abdomen, legs, etc.). The terms "superior" and "inferior" have nothing at all to do with the relative importance of the two veins; rather, the terms refer only to their anatomical positions (the superior vena cava entering the heart from above, and the inferior vena cava entering the heart from below). The heart chamber into which the blood is emptied is the **right atrium** (structure *3* in Fig. 7-3).

Upon filling with blood, the right atrium contracts, pushing the blood downward and into the **right ventricle** (structure *5* in Fig. 7-3). In order to enter the right ventricle, the blood must push upon a valve that separates the two heart chambers; this valve (structure *4* in Fig. 7-3) is called the **tricuspid valve**. The walls of the right ventricle are much more muscular than the walls of the right atrium. This is because the contractions of the right ventricle must push the blood through the entire *pulmonary* circulation, whereas the contractions of the right atrium act only to push the blood through the tricuspid valve. Therefore, contraction of the right ventricle exerts considerable force on the blood and pushes it up and out of the heart through the **pulmonary artery** (structure *8* in Fig. 7-3). In order to enter the pulmonary artery, the blood must push open the second valve of the right side of the heart, namely the **pulmonary semilunar valve** (structure number *7* in Fig. 7-3). The pulmonary artery divides to produce two branches that direct blood to the right and left lungs.

The contraction of the right ventricle is quite vigorous and were it not for the actions of the **papillary muscles** (structure *6* in Fig. 7-3), the upward push of the blood might cause it to regurgitate into the right atrium. This is prevented by the contractions of the papillary muscles which are attached to the walls of the right (and left) ventricle and which pull downward on the tricuspid valve with enough force to exactly counterbalance the upward force of the blood. As a result, the tricuspid valve remains closed and all of the blood of the right atrium is pushed into the pulmonary artery.

Blood returns from the pulmonary circulation to the left side of the heart through the four **pulmonary veins** (structure *9* in Fig. 7-3), two veins carrying blood from the right lung and two from the left lung. This blood enters and fills the **left atrium** (structure *10* in Fig. 7-3). After filling, the left atrium contracts, thereby pushing open the **bicuspid** (or **mitral**) **valve** (structure *11* in Fig. 7-3) and directing the blood into the **left ventricle** (structure *12* in Fig. 7-3). The walls of the left ventricle are the most muscular of all the heart's chambers; this is because the contractions of the left ventricle must push the blood through the entire systemic circulation. Contraction of the left ventricle pushes the blood up and out of the heart through the opened **aortic semilunar valve** (structure *14* in Fig. 7-3) and into the **aortic arch** (structure *15* in Fig. 7-3). At the same time that the left ventricle contracts, so do the papillary muscles attached the inside walls of the left ventricle. These muscles pull down on the bicuspid valve, thereby opposing the upward push exerted by the blood. As a result, the bicuspid valve remains closed and no blood is regurgitated into the left atrium. Branches of the aorta direct the systemic blood to all of the organs of the body.

Figure 7-3

Anatomy of the heart, as seen in a frontal view. *1*, superior vena cava; *2*, inferior vena cava; *3*, right atrium; *4*, tricuspid valve; *5*, right ventricle; *6*, papillary muscle; *7*, pulmonary semilunar valve; *8*, pulmonary artery; *9*, pulmonary veins; *10*, left atrium; *11*, bicuspid valve (also called mitral valve); *12*, left ventricle; *13*, papillary muscle; *14*, aortic semilunar valve; *15*, aortic arch.

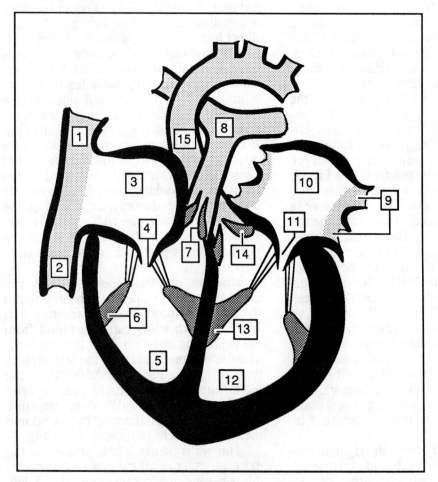

Arteries vs. Veins

It is worth pointing out once again that the terms *artery* and *vein* identify blood vessels on the basis of the *direction* in which they conduct blood through the circulatory system; that is, arteries conduct blood *away from the heart*, whereas veins conduct blood *back to the heart*. Note that blood vessels are *not* identified as arteries or veins on the basis of whether the blood that they contain is "rich" or "poor" in oxygen. Consequently, it should not be a source of confusion for you that the pulmonary artery is poor in oxygen, whereas the pulmonary veins are rich in oxygen.

Timing of Atrial and Ventricular Contractions (Systole and Diastole)

The above discussion of the anatomy of the heart noted the order in which blood travels through the right and left sides of the heart. For convenience, the right side was described first, and then the left side was described. However, it is important to recognize that, in fact, the right and left sides of the heart act synchronously. That is, both the right and left atria fill with blood at the same time and then contract to push the blood into the right and left ventricles at the same time. As the two atria contract, the right and left ventricles are synchronously filled. In a corresponding manner, the right and left ventricles contract to expel blood into the pulmonary artery and the aortic arch at the same time.

The contraction of a heart chamber is called **systole**, and the relaxation that permits the chamber to be filled is called **diastole**. Thus, right and left atrial systole occur at the same time and are accompanied by right and left ventricular diastole. Following this, the right and left ventricles enter systole directing blood into the pulmonary artery and aortic arch. The latter action occurs at the same time as right and left atrial diastole.

Starling's Law of the Heart, Stroke Volume, and Minute Volume

According to **Starling's Law of the Heart**, each of the heart's chambers contracts with sufficient force to expel all of the blood that the chamber contains. Thus, during exercise, when the contractions of skeletal muscle push greater volumes of blood back to the heart, the chambers fill with more blood; this is followed by a corresponding increase in the force of contraction to ensure that all of the blood is expelled.

The volume of blood expelled from the left ventricle with each systole is called the **stroke volume**. In the average person, this amounts to about 70 c.c. of blood. Of course, if the stroke volume is 70 c.c., then each atrial systole also pumps 70 c.c. into the ventricles, and right ventricular systole pumps 70 c.c. of blood into the pulmonary circulation. In other words, whatever volume of blood is pumped from one heart chamber must also be pumped from the others (the atria cannot pump more blood into the ventricles than the ventricles then pump out [which, if it did occur, would violate Starling's Law]). This also implies that equal volumes of blood are pumped into the pulmonary and systemic circulations. During exercise, the stroke volume may increase to 100 c.c. or more.

In a person at rest, the heart goes through its cycle about 70 times per minute (i.e., the resting heart rate is 70 "beats" per minute). If the stroke volume is 70 c.c. and the heart rate is 70 cycles per minute, then in one minute, 4,900 c.c. of blood (i.e., 70 X 70) are pumped from the left ventricle. The volume of blood pumped from the left ventricle in one minute is called the **minute volume**. If, during exercise, the stroke volume increases to 100 c.c. and the heart rate increases to 120 cycles per minute, then the minute volume would be increased to 12,000 c.c. (i.e., 100 X 120). The total volume of blood in the body can be determined and in the average adult is about

5,000 c.c. Thus, if the minute volume at rest is 4,900 c.c., then in one minute's time, a volume of blood equal to the total volume of blood in the body is pumped from the heart.

The Heart Sounds

If in a quiet room, you place your ear against the chest of another person, you will hear certain sounds emanating from the heart; these are called the **heart sounds**. The heart sounds may be heard even more effectively using a **stethoscope**.

There are two heart sounds; the first is called the "**lub**" sound and the second the "**dupp**" sound. These names were chosen because they sound very much like the heart sounds themselves. The sounds are produced by the closures of the heart valves. The lub sound is produced by the closure of the tricuspid and bicuspid (i.e., mitral) valves and occurs at the onset of ventricular systole. The dupp sound is produced by the closure of the pulmonary and aortic semilunar valves and occurs at the end of ventricular systole. It is important to note that neither sound is produced by the "beating" action of the heart *per se* (i.e., you are not hearing the contraction of muscle).

BLOOD PRESSURE

Units of Hydrostatic Pressure

As the blood travels through the body it exerts pressure against the walls of the vessels through which it is flowing. The pressure is called **blood pressure**. Blood pressure is a special form of a more general pressure called **hydrostatic pressure**, which is the pressure exerted by *any* fluid on the walls of the channels or vessels through which the fluid travels.

How pressure changes as the fluid travels forward can be illustrated by considering water from an open faucet traveling through a garden hose that is lying on the ground. If tiny holes are made in the hose along its length, water will not only come out of the end of the hose, but will also spray out of these holes. If all of the holes face upwards, the water will shoot upward to a height that is proportional to the hydrostatic pressure in the hose at the position of the tiny hole. The closer the hole is to the faucet, the greater will be the height reached by the spray.

If narrow glass tubes (columns) are inserted into the tiny openings in the hose (so that the water cannot simply spray out onto the ground [Fig. 7-4]), something different is observed. At each opening, the water rises up in the glass tube until a specific height is reached. At that height, the weight of the column of water inside the glass tube is great enough to exactly resist the upward pressure from below. For holes that are further and further away from the faucet, the height reached is proportionately lower.

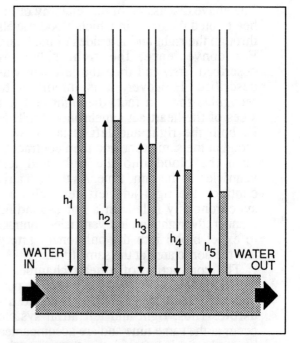

Figure 7-4
Relationship between distance from source of hydrostatic pressure and the magnitude of the pressure at that distance. The hydrostatic pressure is equal to the height the water reaches in each narrow tube (see text).

The amount of hydrostatic pressure present at each point along the hose can be quantitated by measuring the height of the column of water supported in each vertical column (i.e., h1, h2, h3, etc.). For example, if the height of water attained in the second tube from the faucet (i.e., h2) is 18 inches, then the hydrostatic pressure at that point along the hose would be stated as "18 inches of water." If the height attained in the fifth tube is seven inches, then the hydrostatic pressure at that point along the hose would be stated as "7 inches of water."

Usually, hydrostatic pressure values are *not* expressed in "inches of water." Instead, the liquid *mercury* (atomic symbol *Hg*) is used. This is because mercury is much heavier (denser) than water, and so the heights of the columns of mercury sustained by hydrostatic pressure are considerably lower (making measurements more convenient). Since mercury is 13.6 times as dense as water, a hydrostatic pressure of 18 inches of water is the same as a pressure of 1.32 inches of mercury (i.e., 18/13.6).This pressure would be written as *1.32 in. Hg.* (During a T.V. weather report, atmospheric pressure is usually given as about 30 inches of mercury. What this means is that the pressure of the atmosphere pushing down on the surface of the earth is equivalent to that of a column of mercury that is 30 inches high.)

Physiologists usually employ the metric system, and in the case of pressure measurements, millimeters (i.e., mm) are used instead of inches. Since one inch equals 25.4 mm, a hydrostatic pressure of *1.32 in. Hg* is the same as *33.5 mm Hg* (i.e., 1.32 X 25.4).

Measuring Blood Pressure

In a manner similar to the garden hose analogy given above, the pressure exerted by blood on the walls of the blood vessels progressively diminishes the further the blood moves from the heart (i.e., the blood vessels are analogous to the hose, and the heart is analogous to the faucet). The decrease in blood pressure that occurs between the aorta and the superior and inferior vena cava is the result of the resistance presented by the intervening blood vessels. This resistance, called the **peripheral resistance**, is due to friction *within* the blood itself and *between* the blood and vessel walls. The magnitude of the changes in blood pressure that take place as the blood travels through the systemic circulation is shown in Figure 7-5.

Figure 7-5
Changes in blood pressure and blood velocity as blood circulates from the left ventricle through the systemic circulation and back to the right atrium (see text for explanation).

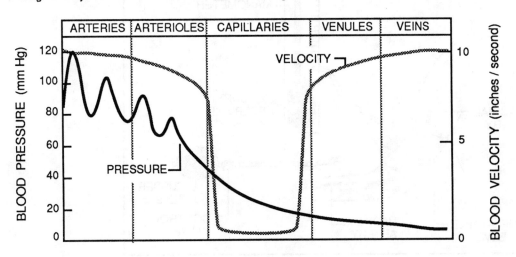

Although there is a continuous fall in pressure from the aorta to the veins, the fall is not uniform (whereas in a garden hose, the fall in hydrostatic pressure would be uniform with increasing distance from the faucet). In a person with normal blood pressure, blood expelled from the left ventricle into the aortic arch raises the pressure in that artery to 120 mm Hg at the peak of ventricular systole; this pressure is called the **systolic blood pressure**. Then, as the left ventricle relaxes and fills with blood, the pressure in the aortic arch falls to about 80 mm Hg; this pressure is called the **diastolic blood pressure**. Thus, the pressure in the aortic arch fluctuates between 120 and 80 mm Hg, depending upon whether the left ventricle is in systole or diastole (Fig. 7-5).

Although blood is pumped into the aortic arch in cycles, it moves forward continuously (not intermittently) in the blood vessels. In the arteries, the flow takes the form of pulsations. This continuous flow is made possible because the walls of the aortic arch are elastic. During systole, the arch is distended, but during diastole the arch recoils, thereby continuing to push the blood forward (the blood cannot back-up because of the aortic semilunar valve).

By the time that the blood reaches the arterioles, the fluctuations in pressure between systole and diastole are no longer discernable. As seen in Figure 7-5, the most dramatic decline in blood pressure occurs across the capillary beds of the body, where the pressure typically falls by 40 mm Hg. By the time that the blood reaches the body's major veins, the pressure is less than 10 mm Hg.

Using a Sphygmomanometer. Certainly, it is not feasible to measure blood pressure in a person in the same way as hydrostatic pressure was measured in the garden hose example given above. That is, one cannot insert tall glass tubes into blood vessels and measure the height of the column of blood supported by the pressure in the blood vessel. (This can be done in the laboratory under experimental conditions and is known as the *direct method* of measuring blood pressure.) Instead, during a physical exam, blood pressure is measured *indirectly* using an instrument known as a **sphygmomanometer** (Fig. 7-6).

Figure 7-6
Using a **sphygmomanometer** to make an indirect measurement of blood pressure. See the text for description of how the sphygmomanometer is used.

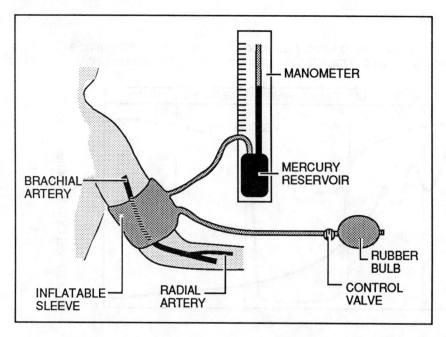

The sphygmomanometer is used to measure the variation in pressure in the **brachial artery** of the upper arm as blood is pumped through this vessel. The procedure takes the following form. The inflatable sleeve of the sphygmomanometer is wrapped around the upper arm of the subject just above the elbow. The bell of a stethoscope is placed on the arm just below the sleeve. With no air in the sleeve and no pressure applied to the upper am, the only sounds heard (if any) will be the smooth flow of blood through the artery. The valve of the rubber bulb is then closed and the bulb repeatedly squeezed in order to pump air into the sleeve. As the sleeve inflates, it applies pressure to the upper arm, further and further occluding the brachial artery. The amount of pressure that is being applied to the arm can be read on the scale of the mercury manometer, which is also connected to the air lines leading to the sleeve.

Typically, air is pumped into the sleeve until the mercury level in the manometer reaches about 200 mm (i.e., the pressure in the sleeve at that point is 200 mm Hg). At 200 mm Hg, the pressure in the sleeve is sufficiently high to squeeze the brachial artery of the upper arm shut, thereby stopping the flow of blood into the lower arm.

With the flow of blood in the brachial artery halted, no sounds are heard with the stethoscope (Fig. 7-7). The valve of the bulb is opened and air slowly released from the sleeve. In a normal person, faint thumping sounds are heard when the mercury level falls to about 120 mm (above 120 mm Hg, no sounds are heard). The faint thumping sounds that are heard at 120 mm Hg are produced by small quantities of blood being pushed through the partially occluded brachial artery *at the peak of ventricular systole* (Fig. 7-7). The pressure at which the first faint sounds are heard is called the **systolic pressure**. As more air leaves the sleeve and the pressure falls further, the thumping sounds get more intense. This is because with each ventricular systole, the volume of blood getting through the brachial artery increases. Eventually, the pressure in the sleeve reaches a low enough value that blood again flows smoothly through the brachial artery. The pressure at which the loud thumping sounds give way to the quieter sounds of even flow is called the **diastolic pressure**. In a normal person, this pressure is about 80 mm Hg.

Figure 7-7
Characteristic sounds are heard emanating from the brachial artery when a sphygmomanometer is used to measure blood pressure. These sounds (and the way that they change) reveal the systolic and diastolic blood pressure.

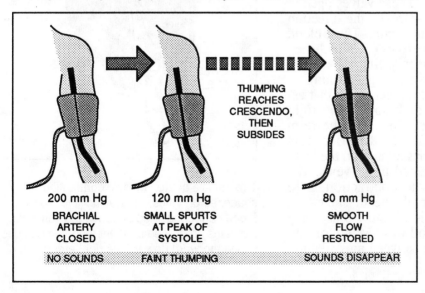

Usually, the systolic and diastolic pressure values are represented as a ratio. Thus, in the above example, the blood pressure would be represented as *120/80*. The numerical difference between the systolic pressure and the diastolic pressure is called the **pulse pressure**. In a normal person, the pulse pressure is 40 mm Hg (i.e., 120 mm Hg minus 80 mm Hg). The pulse pressure represents the pressure variation to which the major arteries are subjected during the cardiac cycle.

The sphygmomanometer makes it possible to obtain an indirect measurement of the fluctuation in blood pressure in a major artery. In stricter terms, the systolic and diastolic blood pressure are defined as follows. The systolic pressure is *the maximum pressure that the blood exerts on the walls of the blood vessel through which the blood flows*, whereas the diastolic blood pressure is the *minimum pressure exerted by the blood on the walls of the blood vessel through which the blood flows*.

Venous Return

As we have seen, the flow of blood in the body's major veins occurs under very little pressure. Indeed, were it not for several accessory mechanisms, the return of blood to the right atrium would be quite inefficient. Venous return is assisted by the presence of *valves* every few inches along the length of a vein. These valves open in one direction only—namely, the direction that leads to the right atrium. Thus, blood cannot back-up in the veins (even in the leg veins where gravity pulls the blood downward, away from the right atrium).

A second important mechanism that assists in the return of blood to the right atrium is the **muscle pump**. The term "muscle pump" refers to the effect that contraction of any musculature in the body has on the flow of blood in the veins. The effect of this muscle contraction is to "pump" blood toward the heart. The mechanism is illustrated in Figure 7-8 and may be explained as follows. Many of the body's veins course between musculature,

and when this musculature contracts, it squeezes the veins. Compression of a short length of a vein forces blood from that length. Although blood is forced in both the forward (i.e., *toward* the heart) and backward (i.e., *away from* the heart) direction, the valves of the veins prevent any net backward flow of the blood. Consequently, the blood can only be displaced in the direction that leads to the right atrium.

So important is the influence of the muscle pump on venous return that in a person who stands motionless, blood has great difficulty returning to the right atrium. Pooling of blood in the lower half of the body due to gravity so deprives the brain of blood (and oxygen) that the person may loose consciousness and collapse.

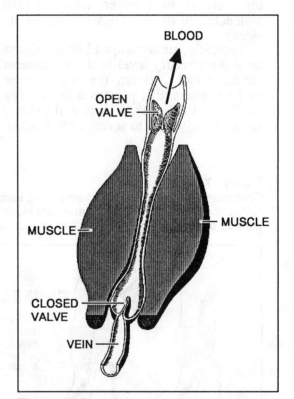

Figure 7-8
The muscle pump. Muscle contraction squeezes on veins, displacing the blood. Blood cannot back-up because the valves in the veins are forced shut. Instead, the blood is pushed forward (toward the right atrium).

Velocity of the Blood

In addition to describing blood pressure changes during circulation, Figure 7-5 shows how the *velocity* (i.e., speed) of the blood changes. Blood pumped into the aortic arch from the left ventricle travels through this major artery at about 10 inches per second. However, as the blood is distributed into the smaller arteries (and then into the arterioles), the velocity of the blood progressively decreases. By the time the blood reaches and flows in the body's capillary networks, its velocity is reduced to about 1/400th of an inch per second.

Changes in blood velocity are analogous to changes in the velocity of any liquid flowing through a closed network of large and small tubes and may be explained as follows (see Fig. 7-9). When a liquid flowing at a given speed in one tube enters a single, narrower tube, the rate of flow increases (Fig. 7-9, *top*). Indeed, the respective rates of flow are *inversely* related to the *cross-sectional areas* of the two tubes. Since a tube's cross-sectional area is a function of the square of the tube's diameter, the rate of flow of a liquid increases four-fold when the liquid passes from one tube into another tube that has half the diameter.

In the case of the circulatory system, however, large arteries branch into a number of narrower arteries, and each of these arteries branches into a number of still narrower arterioles. Unlike the conditions given in the previous paragraph, in the circulatory system a single large tube gives rise to *many* smaller ones (Fig. 7-9, *bottom*). To understand how the velocity of the blood changes as it passes from one vessel into the vessel's many narrower branches, we must consider the *sum* of the cross-sectional areas of all of the branches. In the body, the sum of the cross-sectional areas of the branches of a vessel is greater than the cross-sectional area of the original vessel. Therefore, as blood flows from a major artery into the artery's branches, the blood slows down. As the blood flows from an artery into the many arterioles formed by that artery, the blood again slows down. And, as blood flows from an arteriole into the many capillaries that it forms, the blood's velocity slows still more.

When blood is collected from a capillary network into a venule, the blood's velocity increases again (i.e., the cross-sectional area of the venule is less than the sum of the cross-sectional areas of the capillaries that merge to form that venule). The blood velocity again increases as blood passes from the venules into the veins. Finally, as blood enters the superior and inferior vena cava from the body's smaller veins, it again speeds up, with the result that blood enters the right side of the heart at the same velocity that it left the left side of the heart (i.e., the blood travels in the caval veins at about 10 inches per second.

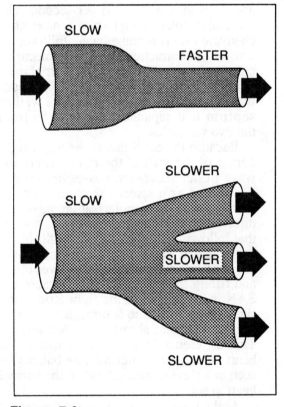

Figure 7-9
Velocity of the blood (see text for explanation).

THE SINO-ATRIAL AND ATRIO-VENTRICULAR NODES

As noted in Chapter 4, cardiac muscle is **myogenic**. That is, heart muscle tissue does not rely on the nervous system for stimulatory impulses. Instead, the electro-chemical events that cause the heart muscle to contract arise within the heart itself. The origin of these electro-chemical events is a small mass of tissue buried in the wall of the right atrium near the superior vena cava's entry point. This small mass of specialized cardiac tissue is called the **sino-atrial node** (usually abbreviated SAN); the SAN is also known as the heart's **pacemaker**.

Approximately once each second (in a person whose heart rate is about 60 cycles per minute), the SAN initiates a wave of depolarization that then spreads across the sarcolemmas (plasma membranes) of the atrial muscle cells (Figure 7-10). The wave of depolarization spreads across the atria at about 20 inches (50 cm) per second. The wave of depolarization (which is an electro-chemical event) is immediately followed by a wave of contraction (which is a mechanical event) spreading through the atrial musculature in the same direction. The depolarization wave does not cross the wall or **septum** that separates the two atria from the two ventricles.

Because the SAN lies in the upper right corner of the wall of the right atrium, the wave of depolarization moves *across* and *downward* as it spreads through the heart tissue. Since the depolarization wave has direction to it, so does the contraction wave that follows. The effect is to drive the blood accumulated in the atria downward, pushing the tricuspid and bicuspid valves open, and filling the two ventricles. Because the SAN is in the wall of the right atrium, the right atrium *begins* to contract a small fraction of a second ahead of the left atrium. This difference is hardly discernable in a heart beating at the normal rate but can be seen in a slow motion picture of the normal heart in action.

At the base of the atria and at the heart's midline is the **atrio-ventricular node** (AVN). This node triggers the contractions of the two ventricles. When atrial depolarization reaches the AVN, the AVN initiates a wave of depolarization that travels down two bundles of cardiac fibers on either side of the wall that separates the right and left ventricles; these structures are the **Bundles of His**. At the *apex* of the heart, the Bundles of His give rise to individual **Purkinje fibers** that radiate upwards terminating in various regions of the ventricular musculature. When the wave of depolarization carriedby the Bundles of His and Purkinje fibers reaches the ends of the Purkinje fibers, the membranes of adjacent cardiac muscle cells are depolarized and the cells contract. Depolarization of the Bundles of His and Purkinje fibers proceeds at about 200 inches (500 cm) per second.

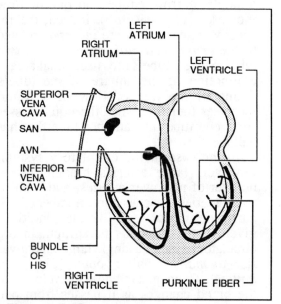

Figure 7-10

The sino-atrial node (SAN), atrio-ventricular node (AVN), Bundles of His, and Purkinje fibers of the heart. In each cardiac cycle, a wave of depolarization spreads across the atria from the SAN. This wave, which is followed by atrial systole, activates the AVN. The AVN initiates a second wave that reaches the ventricular musculature via the Bundles of His and Purkinje fibers and is followed by ventricular systole.

Just as in the case of the atria, contraction of the ventricles has direction. The depolarization wave first reaches the ventricular muscle cells near the heart's apex because the Purkinje fibers terminating in this region are the shortest (i.e., the depolarization wave travels the length of a short fiber in less time than it travels the length of a long fiber). The last ventricular cells to be depolarized are those near the atrio-ventricular septum, because this area is reached by the longest Purkinje fibers. Thus, contraction of the ventricles begins at the apex of the heart and spreads upward toward the atria, pushing the blood ahead of it. Blood cannot be regurgitated into the two atria because the tricuspid and bicuspid valves are kept closed. Therefore, the blood passes into the pulmonary artery and aortic arch, pushing the pulmonary and aortic semilunar valves open as the blood advances.

As noted earlier, the AVN lies at the heart's midline; additionally, there is symmetry in the organization of the left and right Bundles of His and Purkinje fibers of the two ventricles. Consequently, the contractions of each ventricle begin (and end) at the same time. As already noted, this is not so in the case of the atria. Because the SAN is located in the wall of the right atrium, right atrial systole begins (and ends) slightly ahead of left atrial systole.

THE ELECTROCARDIOGRAPH

The changes in electrical activity of the heart that precede atrial and ventricular systole are conducted through other tissues of the body to the body's surface and may be detected and recorded using an instrument known as an **electrocardiograph**. The recording that is obtained is called an **electrocardiogram** (usually abbreviated **EKG** or **ECG**).

An EKG of the heart's activity is obtained by attaching the sensing electrodes of the electrocardiograph to the skin of the arms, legs, and/or chest. A permanent recording may be obtained using graph paper, or the tracing may be displayed transiently on an oscilloscope or television monitor. In either case, what is seen is the relationship between elapsing time (usually measured in seconds) and the magnitude of the electrical changes taking place in the heart tissues (usually measured in millivolts [i.e., thousandths of a volt]). A typical tracing for a normal heart is shown in Figure 7-11.

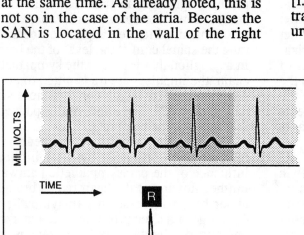

Figure 7-11
In the upper part of the diagram, four successive cycles of an electrocardiograph tracing are shown. In the enlarged view of one of these cycles shown below, the P, QRS, and T waves are identified.

The electrocardiogram shows three successive "waves" (rises and falls) for each cycle of heart activity. The first wave is the **P-wave**. This wave is a measure of the electrical changes in the heart that accompany the *depolarization* of the sino-atrial node and the atrial musculature. It is important to remember that this wave precedes the contractions of the two atria. The P-wave begins at the sino-atrial node and then spreads across the atrial muscle tissue.

The P-wave is followed by the **QRS-wave (or QRS-complex)**. Point Q represents the onset of the wave, point R is the wave's peak, and point S is the end of the wave. The QRS-wave corresponds to the *depolarization* of the atrio-ventricular node, the Bundles of His, the Purkinje fibers, and the ventricular musculature. Notice how much larger the QRS-wave is than the P-wave. The QRS-wave is greater because so much more tissue is depolarized. The QRS-wave is followed by ventricular systole.

The third (and last) wave of the electrocardiograph tracing is the **T-wave**. This wave represents the *repolarization* of the ventricular tissues. To be sure, the atrial tissues are also repolarized before the next cardiac cycle begins; however, no distinct wave is produced during atrial repolarization. This is because the repolarization of the atria takes place at the same time as the depolarization of the ventricles. Since the depolarization of the ventricles represents such a dramatic electrical change, it masks the effects of atrial repolarization.

It is important to note that the EKG tracing is a measure of electrical changes in the heart tissues and not mechanical changes. That is, the EKG is not a measure of the timing or force of heart muscle contraction. Rather, it is a measure of the electrical events in the heart that precede muscle contraction.

CONTROL OF HEART ACTIVITY

Although heart tissue is myogenic, the rate and force with which the heart beats can be altered by the nervous system. Regulation of the heart is effected through sympathetic and parasympathetic pathways that link the brain and spinal cord with the SAN (and AVN). Increases in the flow of nerve impulses to the heart over sympathetic pathways serve to increase heart rate; whereas increases in the flow of nerve impulses to the heart over parasympathetic pathways serve to decrease heart rate. Usually, increased activity of one pathway is accompanied by decreased activity of the other pathway, a phenomenon known as **reciprocal innervation**.

The parasympathetic pathway begins in a center in the medulla of the brainstem called the **cardioinhibitory center** (CIC). Motor impulses are conducted from the CIC to the heart over pre-ganglionic fibers that are a part of the Vagus nerve (i.e., cranial nerve number 10). In the wall of the right atrium is the terminal ganglion in which the pre-ganglionic fibers synapse with short post-ganglionic fibers. The post-ganglionic fibers terminate in the SAN.

The center for the sympathetic pathway also resides in the medulla and is called the **cardioaccelerator center** (CAC). Impulses originating in this center are conducted down the spinal cord, where they are transmitted to pre-ganglionic fibers exiting the spinal cord at the level of the heart. In a ganglion that is part of the sympathetic chain, the impulses are transmitted to post-ganglionic fibers which conduct these signals to the SAN. The effect of this pathway is to increase the heart rate.

Under resting conditions, the heart rate is determined primarily by the inhibitory influence of the parasympathetic pathway. Further slowing of the heart rate is brought about by an increase in parasympathetic discharge and a reciprocal decrease in sympathetic discharge. The heart rate is increased by increasing the sympathetic impulses and reciprocally decreasing the parasympathetic impulses.

The respective activities of the CIC and CAC are determined by a number of factors including (1) the nature of impulses reaching these centers from higher regions of the brain (e.g., the cerebrum), (2) impulses reaching these centers from interoceptors in the aorta and carotid arteries, and (3) the chemical composition of the blood flowing through the medulla itself.

The effects of the cerebrum on heart rate should already be quite familiar to you. Watching a frightening movie or the exciting finish to a sports event can cause an increase in heart rate and blood pressure. These effects are due to the influence of the higher brain centers of the cerebrum upon the cardiovascular centers of the brainstem.

Changes in heart rate and blood pressure can also be induced reflexively through the actions of interoceptors located in the walls of blood vessels near the heart. As shown in Figure 7-12, blood entering the aortic arch from the left ventricle is apportioned among a number of arterial branches. The first branches are the **coronary arteries**, which exit the aortic arch just beyond the aortic semilunar valve. The coronary arteries carry blood into the walls of the heart. (The heart itself is, therefore, the first organ whose tissues receive fresh, oxygen-rich systemic blood). The second branch of the aorta is the **innominate artery** (also

known as the **brachiocephalic artery**). This is a short artery that soon divides into two branches. One of these branches is the **right subclavian artery**, which directs blood to the chest and shoulder regions on the right side of the body and also into the right arm. The other branch is the **right common carotid artery**, which directs blood up through the neck and into the head. The third branch of the aortic arch is the **left common carotid artery**. (Note that unlike its right counterpart, the left common carotid artery branches directly from the aortic arch; there is no innominate [or brachiocephalic] artery on the left side of the body.) Like the right common carotid artery, the left common carotid artery courses upward through the neck. The fourth branch of the aortic arch is the **left subclavian artery** (again note that unlike its right counterpart, the left subclavian artery branches directly from the aortic arch). The left subclavian artery directs blood to the chest and shoulder regions on the left side of the body and also into the left arm.

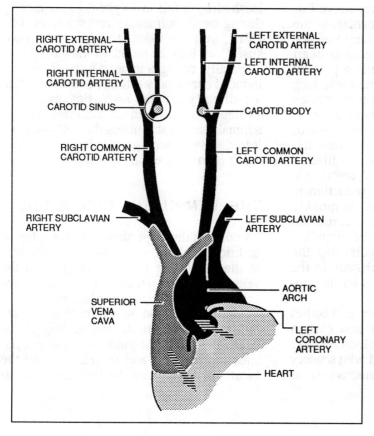

Figure 7-12
The walls of the major arteries and veins near the heart contain interoceptors that sense blood pressure. Nerve impulses initiated by these interoceptors serve to reflexively regulate heart rate and blood pressure. See the text for discussion.

RIGHT EXTERNAL CAROTID ARTERY
RIGHT INTERNAL CAROTID ARTERY
CAROTID SINUS
RIGHT COMMON CAROTID ARTERY
RIGHT SUBCLAVIAN ARTERY
LEFT EXTERNAL CAROTID ARTERY
LEFT INTERNAL CAROTID ARTERY
CAROTID BODY
LEFT COMMON CAROTID ARTERY
LEFT SUBCLAVIAN ARTERY
AORTIC ARCH
SUPERIOR VENA CAVA
LEFT CORONARY ARTERY
HEART

Especially important insofar as the regulation of heart rate and blood pressure are concerned are the left and right common carotid arteries of the neck and their respective external and internal branches. Each common carotid artery gives rise to two branches. One of these branches, the **external carotid artery**, carries blood upward from the neck into the superficial regions of the head (facial muscles, scalp, etc.), whereas the other branch, the **internal carotid artery**, delivers blood to the deeper tissues of the neck and head, including the brain. At the branch points, there is a dilation of each internal carotid artery; these dilations are the **carotid sinuses** (Fig. 7-12). Among the cells that make up each carotid sinus are *baroreceptors* that sense blood pressure. Buried in the wall of each carotid sinus are clusters of *chemoreceptors* that sense the concentrations of oxygen and carbon dioxide dissolved in the blood. Similar baroreceptors and chemoreceptors are located in the walls of the aortic arch.

The pressure receptors of the carotid sinuses and aortic arch are associated with sensory nerve fibers that carry nerve impulses to the cardiovascular centers in the medulla. Suppose that something you see or hear causes a sudden increase in heart rate. When this occurs, there is a parallel increase in blood pressure in the aortic arch and carotid arteries. This, in turn, causes the frequency of discharge from the receptors to increase. The increased frequency of sensory impulses reaching the medulla reflexively causes a widespread dilation (relaxation) of the walls of arterioles, bringing about an immediate reduction in blood pressure. Usually, this is quickly followed by a decrease in sympathetic and an increase in parasympathetic impulses sent to the SAN, thereby decreasing the heart rate. A fall in blood pressure in the aortic arch and carotid arteries has the opposite effects.

The chemoreceptors of the carotid bodies sense the levels of oxygen and carbon dioxide dissolved in the blood. These chemoreceptors are associated with sensory nerve fibers that carry a continuous flow of impulses to the medulla. When the oxygen level falls and/or the carbon dioxide level rises, there is an increase in the frequency of these impulses and this triggers an increase in heart rate so that larger amounts of blood are sent through the pulmonary circulation. An increase in the oxygen content of the blood (or a decrease in the blood's carbon dioxide) has the opposite effect.

The amounts of oxygen and carbon dioxide dissolved in the blood that flows through the medulla also act to influence heart rate. The quantities of these two gases in the blood vary inversely; that is, a decrease in blood oxygen (usually due to increased body activity) is accompanied by an increase in carbon dioxide, and vice-versa. The CAC and CIC of the medulla sense changes in the levels of these dissolved gases and trigger the necessary changes in heart rate.

It is interesting that the cardiovascular centers of the brain are much more sensitive to carbon dioxide than they are to oxygen. That is to say, a much earlier and rapid response is triggered by a rise in the level of carbon dioxide in the blood than is unleashed by a fall in oxygen. (This is why during cardiopulmonary resuscitation [i.e., CPR], you are asked to blow your own "used" air into the mouth of the victim.)

Finally, changes in heart rate can also be induced chemically through substances that act directly on the SAN. For example, epinephrine (a hormone secreted by the adrenal glands) stimulates the SAN causing it to discharge with greater frequency. The result is an increase in heart rate.

THE LYMPHATIC CIRCULATION

As noted at the beginning of this chapter, in addition to blood, a second fluid circulates in the body; this fluid is **lymph** and the vessels through which it circulates comprise the **lymphatic circulation**.

Unlike the blood, which travels in a *circular* route beginning and ending at the heart, the lymph originates in various tissues of the body and travels *toward* the heart only. The process that gives rise to lymph is illustrated in Figure 7-13. Blood entering the capillary beds of a tissue is the source of nutrients, raw materials, oxygen, and other life-sustaining substances. Quantities of these materials, together with water, are filtered from the bloodstream at the arterial ends of the capillaries and enter the **tissue spaces** that surround the tissue cells. The fluid that fills these spaces and bathes the tissue cells is called **tissue fluid**. No red blood cells are filtered from the blood, so that tissue fluid is colorless. Cells draw upon the tissue fluid for their metabolic needs and also empty their wastes and secretions into this fluid.

Much of the material emptied into the tissue spaces from the tissue cells enters the blood near the venous ends of the capillaries. However, some of the tissue fluid enters more porous vessels that arise blindly within the tissue; these blind-ended vessels are the **lymphatics**. Once within the lymphatic vessels, the fluid is now called lymph.

Lymph is drained from a tissue by successively larger lymphatics that ultimately convey and empty the lymph into specific veins. The unidirectional flow of lymph within the lymphatics, which is maintained despite the absence of a pump, relies on the contractions of visceral and skeletal muscles. These contractions squeeze upon the lymphatics that course through the tissue. The lymphatics contain a rich supply of valves that open only in the direction that leads to the right side of the heart.

All lymphatics ultimately converge upon two major lymphatic vessels that empty the lymph into the blood of the **brachiocephalic veins** (Fig. 7-14). These vessels are the **right lymphatic duct** and the **thoracic duct** (the latter is also known as the left lymphatic duct). Of the two, it is the thoracic duct that is responsible for carrying by far the largest amount of lymph into the bloodstream (many more lymphatic vessels converge upon the thoracic duct than converge upon the right lymphatic duct). Once the lymph enters the brachiocephalic veins, it is rapidly mixed with the blood. The brachiocephalic veins merge to form the superior vena cava, which then empties the mixture of blood and lymph into the right atrium.

Figure 7-13

Water, nutrients, raw materials, and oxygen pass from the bloodstream into the tissue spaces as blood circulates through the capillary beds of the body. Much of the waste produced by tissue metabolism and many of the tissue's secretions enter the bloodstream near the venous ends of the capillary beds and are carried back to the right side of the heart. Additional tissue waste and other tissue secretions (e.g., hormones) pass from the tissue space into blind-ended lymphatic vessels, thereby forming lymph. The lymph slowly progresses toward and eventually enters the blood of the major veins near the heart.

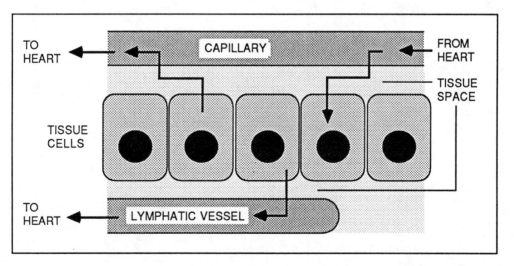

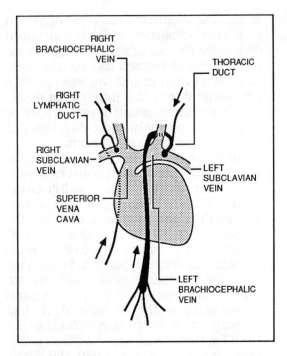

Figure 7-14
Lymph produced throughout the body is eventually collected by the **right lymphatic duct** and the **thoracic duct**. These major lymphatic vessels empty their lymph into the blood of the right and left **brachiocephalic veins** (near their junctions with the subclavian and jugular veins).

The lymphatic circulation is characterized by large numbers of lymph nodes (Fig. 7-15) that interrupt the flow of lymph. Sev-

eral lymphatic vessels (called **afferent lymphatics**) may enter a single node, but the node is usually drained by a single **efferent lymphatic**. The nodes are large masses of lymphoid tissue that are perforated by sinuses. The lymph is filtered as it flows through these sinuses, foreign matter and particulate debris being removed and degraded by the nodal tissue.

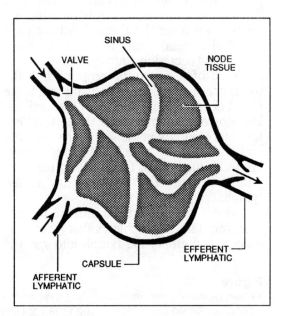

Figure 7-15
Organization of a lymph node. The arrows show the direction of lymph flow into and out of the node. See text for explanation.

CHAPTER 8

THE BLOOD

In this chapter, we will be concerned with the composition and physiological functions of the blood. Although we will consider each function in some detail, the various functions of the blood may briefly be summarized as:

1. to transport oxygen to the tissues of the body,
2. to carry away waste carbon dioxide,
3. to transport absorbed nutrients and raw materials to the tissues of the body,
4. to carry away from the tissues its various non-gaseous metabolic wastes,
5. to shuttle substances (e.g., hormones) from one tissue to another,
6. to protect the body against infection, and
7. to distribute the body heat .

COMPOSITION OF THE BLOOD

Since under normal conditions a sample of blood removed from the body undergoes a transformation from a liquid state to a solid state (i.e., it **coagulates**), the only simple way to study the composition of the blood is to collect a sample into a container that has been coated with a chemical substance that prevents coagulation (i.e., an **anticoagulant**). If this is done and the "whole blood" is allowed to stand upright for several hours, the blood will separate into two major phases (Figure 8-1). The lower phase is intensely red in color and is called

the **formed element** phase; as we will see later, this phase consists principally of red blood cells or erythrocytes. The upper phase is a clear, yellow fluid and represents the medium in which the formed elements were previously suspended; this phase is called the **plasma.**

A number of substances can act as anticoagulants. The body's natural anticoagulant is **heparin**; its function is to prevent blood from coagulating during its circulation through the body. Other common anticoagulants include *sodium citrate*, *sodium oxalate*, and *ethylenediamine tetraacetic acid* (usually abbreviated *EDTA*).

In a normal person, the formed elements represent about 45% of the total volume of a sample of whole blood; the other 55% is represented by the plasma (see Fig. 8-1). This ratio (i.e., 45% or 0.45) is called the **hematocrit** and is a convenient, quickly-measured, and simple diagnostic indicator of the condition of the blood. For example, if a person's hematocrit were 30% (or 0.30), this would be an indication of the existence of an anemia, there being too few (red) blood cells in the blood.

When fresh, whole blood is collected into a container that lacks anticoagulant, the blood quickly undergoes a transformation from a liquid to a gelatinous solid (Fig. 8-1). The solid is called a **blood clot**. If the clot is allowed to stand for several hours, it shrinks or **retracts**, exuding a clear fluid, and slowly forming a smaller and more

119

solid mass. Eventually, the retracted clot settles to the bottom of the container, leaving the clear fluid above. The clear fluid exuded from the retracted clot looks very much like plasma but it is called **serum**. There is an important difference between the compositions of blood plasma and blood serum; namely, plasma contains the chemicals needed in order to cause the coagulation of blood, whereas serum lacks these substances because they were consumed during the formation of the clot.

THE FORMED ELEMENTS

The formed element phase of a blood sample represents all of the solids that settle to the bottom of the container when whole blood is collected into anticoagulant. These solids include a number of different cells and particles. The predominant constituents of the formed element phase are the red blood cells or **erythrocytes**. These contain the red protein **hemoglobin**, which transports oxygen in the blood and also gives the blood its red color. In a normal person, there are about 5,000,000,000 (i.e., five billion) erythrocytes in each c.c. (cubic centimeter) of whole blood.

Also present among the formed elements but much fewer in number are the white blood cells or **leukocytes**. In a normal person, there are about 5,000,000 (i.e., five million) leukocytes in each c.c. (cubic centimeter) of whole blood. As we will see later, the leukocytes play a variety of roles in the body. Although most leukocytes are larger than erythrocytes, they are considerably less dense. Thus, being lighter, they settle from whole blood more slowly than the erythrocytes and form a thin layer called the **buffy coat** at the top of the formed element phase (see Fig. 8-1).

Finally, the formed element phase also includes a large number (i.e., 250,000,000/c.c.) of non-cellular particles called **platelets** or **thrombocytes**. These play an important role during the coagulation of blood that occurs following injury. The platelets are much smaller than the red and white blood cells and settle very slowly onto the top of the buffy coat.

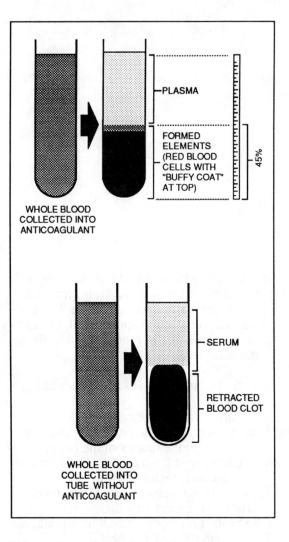

Figure 8-1
Top: When whole blood is collected in anticoagulant, it soon settles into the formed element phase and overlying plasma. *Bottom:* If collected in the absence of anticoagulant, the blood clots and retracts, the retracted clot settling to the bottom of the container; the overlying clear liquid is called serum.

THE BLOOD PLASMA

The blood plasma is the fluid in which the formed elements are suspended. However, plasma should not be thought of as an inert liquid playing a passive physiological role. Despite being mainly water, the plasma is rich in dissolved proteins that serve a variety of important physiological roles in the body and is the medium in which such vital substances as vitamins, sugars, amino acids, and hormones are transported from one tissue site to another. We will look more closely at the various functions of the plasma later in the chapter, but at this point let's examine the red blood cells, white blood cells, and platelets in greater detail.

RED BLOOD CELLS (ERYTHROCYTES)

Shape and Numbers of Red Cells

The red blood cells are the most abundant of the formed elements (5,000,000,000/ c.c. whole blood) and are also among the simplest and smallest of the cells in the body (Figure 8-2). Red blood cells have no nucleus and also lack most of the other intracellular components that characterize typical human cells (e.g., mitochondria, Golgi bodies, endoplasmic reticulum, ribosomes, etc.). The interior of the cells is packed with molecules of the conjugated protein hemoglobin which serves to transport oxygen from the lungs to the body's tissues. Indeed, the amount of hemoglobin present in red cells is so great that it is necessarily arranged in an orderly manner with the result that it appears almost crystalline (i.e., "para-crystalline"). The para-crystalline organization of hemoglobin in red blood cells is believed to contribute to the peculiar shape of the cell which is that of a **biconcave disk** (Fig. 8-2). Because erythrocytes lack a cell nucleus and other components, they are capable of only limited metabolism and are certainly unable to grow and divide. Thus, red blood cells have the rather limited life-span of about 120 days.

Erythrocytes of individuals with abnormal hemoglobins (hemoglobins containing one or more amino acid substitutions) usually lack the normal biconcave shape. The most notorious example is the "sickle" or crescent shape of erythrocytes in people with **sickle-cell anemia**. In this geneti-

cally-determined disease, a single amino acid substitution occurs in the beta globin polypeptide chains of the hemoglobin. Under conditions of oxygen deprivation or shortage, the hemoglobin molecules aggregate to form long fibers and these progressively group together into bundles. The bundles of hemoglobin molecules make the cells less flexible and also deform the cells, causing their sickling.

The oxygen transported by erythrocytes enters and leaves the cells by diffusion, and the biconcave shape of the cell facilitates oxygen movement by increasing the surface area-to-volume ratio of the cell. The disk shape of the erythrocyte also induces the formation of **rouleaux** or long stacks of cells, with the result that far greater numbers of cells can pass through the narrow capillaries when arranged in such a regimented manner than when the cells are freely and independently suspended (Figure 8-2).

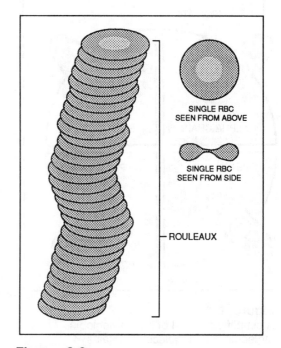

SINGLE RBC
SEEN FROM ABOVE

SINGLE RBC
SEEN FROM SIDE

ROULEAUX

Figure 8-2.
Red blood cells lack nuclei and other intracelular components and have the shape of biconcave disks; this shape lends itself to an orderly stacking of cells, thereby forming **rouleaux**.

Hemoglobin

Hemoglobin is a *conjugated* protein; that is, it is a protein that contains some non-protein portions. The function of hemoglobin is to transport oxygen, and therefore hemoglobin molecules can exist in either of two states: *oxygenated* and *unoxygenated*. The most common type of human hemoglobin is called hemoglobin **A** (A = "adult") and is usually abbreviated **HbA**. HbA is a relatively small conjugated protein and has a molecular weight of 64,500 (Fig. 8-3). The protein portion of the molecule, called **globin**, has 574 amino acids distributed among four polypeptide chains. The four globin chains consist of two identical pairs: two *alpha* chains (141 amino acids each) and two *beta* chains (146 amino acids each).

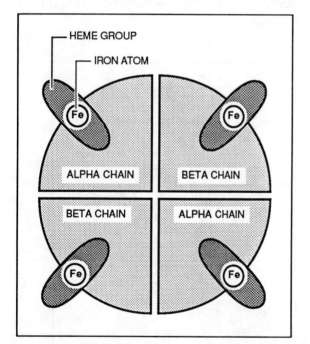

Figure 8-3
A simplified model of the hemoglobin molecule. Each molecule contains 4 globin chains (polypeptides), a pair of *alpha* chains and a pair of *beta* chains. Each globin chain contains one heme group, whose central iron atom (Fe) can bind one molecule of oxygen (i.e., O_2, two atoms of oxygen).

The non-protein portions of each hemoglobin molecule consist of four **heme groups**, one heme group associated with each of the four globin chains. At the center of each heme group is an all-important iron atom. It is this iron atom with which *molecular* oxygen (i.e., O_2) combines in the lungs to form oxygenated hemoglobin and from which oxygen is released in other tissues of the body. The reactions may be summarized thusly:

In the lungs: $Hb + 4O_2 \rightarrow Hb(O_2)_4$

In other tissues: $Hb(O_2)_4 \rightarrow Hb + 4O_2$

Note that each iron atom (or heme group) combines with a *molecule* of oxygen, so that altogether 8 atoms of oxygen can be transported by one molecule of hemoglobin.

In adults, HbA is the most abundant hemoglobin and accounts for 98% of all hemoglobin present. A second adult hemoglobin, called **HbA2**, accounts for the remaining 2%. The differences between HbA and HbA2 occur in the beta globin chains. Still other kinds of hemoglobin molecules are present in the blood at other stages of development. In the very early stages of embryonic development, a hemoglobin called **Gower-1** appears. The next hemoglobin to appear is called **Gower-2** and this is quickly followed by the appearance of **Hb Portland**. Both Gower-2 and Portland are soon replaced by **HbF** or **fetal hemoglobin**. HbF predominates in the blood through the remainder of fetal development. Beginning soon after the twelfth week of fetal development, HbA appears and progressively increases in quantity. Shortly before birth, HbA2 appears. By about six months after birth, little if any HbF can be found in the blood. At this time, about 98% of the hemoglobin is HbA and the remaining 2% is HbA2, and in a normal person, this ratio persists through adult life.

Erythropoiesis

Red blood cells are produced in the **bone**

marrow, a soft tissue that fills small cavities at the centers of many bones. Blood flows into the marrow via small arterioles that enter the bone through tiny surface perforations. The vessels terminate within the marrow, with the result that the blood is free to percolate through the marrow tissue. The narrow spaces through which the blood passes are called the **marrow sinuses**. Having traveled through the sinuses, the blood is then collected by small venules that direct the blood back out of the bones. While percolating through the marrow sinuses, new red and white blood cells are added to the blood.

The development of red blood cells in the bone marrow is called **erythropoiesis** and is one of the most striking examples of cell specialization occurring in the body. Erythropoiesis begins with a *pluripotent* **stem cell** that gives rise in 4-7 days to mature, hemoglobin-filled erythrocytes. While the specialization of the red cell is accompanied by the production of large amounts of hemoglobin, it is also accompanied by the loss of nearly all internal cell structures and many physio-logical properties. Mature red blood cells lack nuclei, mitochondria, endoplasmic reticulum, Golgi bodies, ribosomes, and most other typical cell organelles.

As noted earlier in the chapter, the mature erythrocyte is a relatively simple cell, bounded at its surface by a plasma membrane and containing internally a highly concentrated, para-crystalline array of hemoglobin molecules that are used for oxygen transport. This highly specialized state is attained during the course of a number of developmental stages that are depicted in Figure 8-4. The development begins with the pluripotent stem cells, that is cells capable of producing leukocytes (white blood cells) as well as erythrocytes. When stem cells undergo division, one of the daughter cells remains undifferentiated and pluripotent, so that depletion of marrow stem cells does not take place. If the other daughter cell produced by stem cell division gives rise to erythrocytes, then that daughter cell is a **pro-erythroblast** (if it gives rise to leukocytes, it is called a **pro-leukoblast**). Hemoglobin synthesis begins at the pro-erythroblast stage.

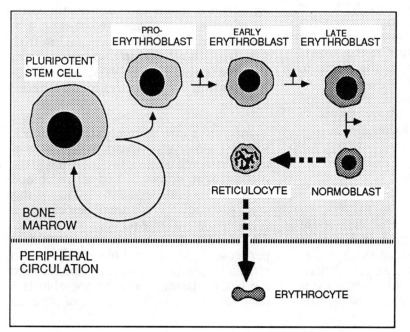

Figure 8-4 Erythropoiesis (the progressive development of mature erythrocytes from stem cells). For simplicity, beginning at the pro-erythroblast stage, only one of the two daughter cells produced by cell division is shown. At the reticulocyte stage, the cells pass from the bone marrow into the peripheral bloodstream.

With very little intervening cell growth, the pro-erythroblast soon divides into two smaller cells called **erythroblasts**. Once this stage is reached, hemoglobin synthesis becomes the dominant aspect of cell metabolism. Again, little cell growth precedes the division of erythroblasts into **normoblasts**. In normoblasts, hemoglobin synthesis reaches its peak. Shortly before the completion of hemoglobin synthesis, each normoblast begins to loose its intracellular structures, most notably the nucleus, and the cell is transformed into a **reticulocyte**. The term reticulocyte refers to the reticulum (i.e., lacy network) that can be seen in this cell when it is stained for microscopy. The reticulum represents a precipitate of the vestiges of intracellular structures. Reticulocytes continue to synthesize hemoglobin for several hours, but afterwards all protein synthesis comes to a permanent halt. It is at the reticulocyte stage that the maturing red blood cell is released from the bone marrow into the circulating blood. Within a few hours, the reticulocyte looses its reticulum and takes on the characteristic biconcave shape of the mature erythrocyte.

Although all protein synthesis comes to a halt during reticulocyte maturation, a red cell retains a limited metabolic capacity after release from the bone marrow. What limited metabolism is retained by the mature cell serves to sustain it during its 120-day and 700-mile journey through the circulatory system. We will return to the subject of erythropoiesis later in the book in connection with our discussion of hormones, for erythropoiesis is regulated by the sceretion of a hormone (called **erythropoietin**) by the kidneys.

Rate of Red Blood Cell Production

Since in the average, normal adult there are 5 billion red blood cells in each c.c. of blood and these cells have an average lifespan of 120 days, a few simple calculations quickly reveal the rate at which erythrocytes are released from the bone marrow into the circulating blood. About 8% of the body weight is represented by blood; therefore, a person weighing 78 kilograms or 172 lb has 6.24 kilograms or about 6.3 liters (i.e., 6,300 c.c.) of blood in his circulation. If each c.c. of blood contains 5 billion red cells, then altogether there would be 31.5×10^{12} (i.e., 31,500,000,000,000) circulating erythrocytes. Since all of these are replaced during an interval of 120 days or 10.4×10^6 seconds, this means that 3.03×10^6 red blood cells reach full maturity and are released into the bloodstream each second. In other words, in a typical adult the differentiation and maturation of 3 million erythrocytes is completed *each second*.

Obviously, an appreciable amount of the body's metabolic energy as well as chemical resources are continuously consumed to support erythropoiesis. This is in stark contrast with other highly differentiated cells, such as those of muscle and nerve, whose proliferation ceases shortly after birth.

Elimination of Old Red Blood Cells

If 3,000,000 new red blood cells are released into the circulating blood each second, then it stands to reason that a corresponding number of aged red cells must be lost each second. Old red cells are lost from the circulating blood in two ways: (1) by **intravascular hemolysis**, and (2) **sequestration** by the **reticuloendothelial system**.

Intravascular Hemolysis. By intravascular hemolysis is meant the breakage of a red blood cell while it is circulating in the bloodstream. About 5% of all red blood cells lost from the circulation are lost in this way. Although 5% is a small percentage, it is nonetheless a large number of cells (e.g., 5% of 3,000,000 per second is 150,000 per second). When a red blood cell lyses in circulation, its hemoglobin is "spilled" into the blood plasma. Since hemoglobin is a fairly small protein, it is capable of filtration in the kidneys, where under the acid conditions that exist in that organ, the hemoglobin would form a precipitate. Such filtration and precipitation of free hemoglobin could do serious injury to the kidney tissue. This is prevented in the following way. Circulating in the blood plasma is another protein called **haptoglobin** which combines with hemoglobin to form a non-filterable and non-precipitating complex that is later removed from the blood by the reticuloendothelial tissues (see below).

Sequestration by the Reticulo-endothelial System. The reticuloendo-thelial system (or **RES**) is a collection of tissues that seek out (i.e., "sequester") aged red cells and remove them from the circulating blood. Tissues of the RES are found in the **bone marrow, liver,** and **spleen** and function similarly in all of these organs. The principal mechanism for seques-

tering aged red cells is shown in Fig. 8-5.

RES tissue forms sinuses in which the blood directly contacts the surrounding tissue cells (i.e., the are no capillary walls separating the blood from the surrounding tissue). Unlike young red cells which readily pass through the sinuses, aged red cells encountering the sinus walls become attached to the surfaces of the cells lining the sinus. These cells are called **macrophages** and having adsorbed an aged red cell, they begin to engulf the cell in a surface pocket. The mechanism is called **phagocytosis** and is a special form of a more general phenomenon called **endocytosis**. Eventually, the red cell is completely enveloped in a small chamber or **vacuole** within the macrophage. The vacuole soon fuses with smaller vesicles (called **lysosomes**) whose rich complement of digestive enzymes carries out the complete degradation of the entrapped red cell.

Figure 8-5
Sequestration of aged red blood cells by phagocytic cells of the reticulo-endothelial system (RES). The numbers 1 through 7 show the progressive phases through which an aged cell passes as it is engulfed and destroyed by the RES. Young red blood cells pass uninterrupted through the sinuses of the RES.

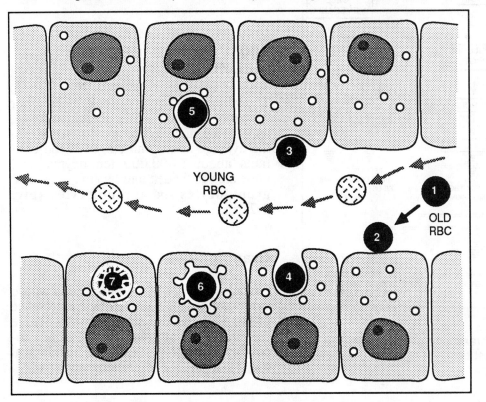

The body expends considerable effort in the production and destruction of red blood cells. The degradation of red blood cells produces a variety of end-products, many of which are re-used or "recycled" by the body (Fig. 8-6). Since the bulk of a red cell is represented by its plasma membrane and enclosed hemoglobin, we may limit our consideration of red cell destruction to the fates of these two materials. Degradation of the plasma membranes of red cells yields two major products: the membrane lipids and the membrane proteins. Both of these are conserved by the body and re-used. Prior to its recycling within the body, the protein of the red cell membrane is broken down to yield the individual amino acids of which the proteins were composed. These amino acids are used in the synthesis of new tissue protein.

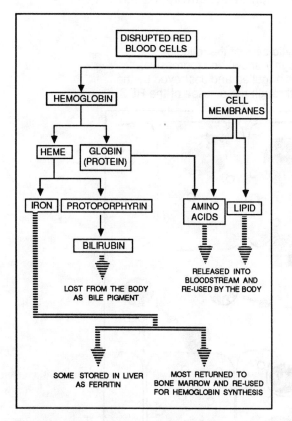

Figure 8-6
Fates of the breakdown products released during the destruction of aged red blood cells by the RES.

The initial stage of hemoglobin breakdown separates the molecule into its protein component (i.e., *globin*) and the iron-containing *heme*. Like the protein of the plasma membrane, globin is broken down to individual amino acids which are re-used for new protein synthesis. Heme is separated into two components: iron and **protoporphyrin**. The iron released from heme is re-used by the body. Since most of the body's iron is present in hemoglobin, the re-utilization of iron implies that most of it will once again be incorporated into the hemoglobin of newly-developing red blood cells. Iron released during the breakdown of hemoglobin is either stored as an iron-rich conjugated protein called **ferritin** or is transported in the blood to the erythropoietic tissues by a plasma protein called **transferrin**.

Of all of the breakdown products of hemoglobin, only the protoporphyrin is excreted by the body. The protoporphyrin is lost from the body either as the **bile pigments** (*bilirubin* and *biliverdin*) or as **urobilinogen**. These substances are responsible for the color of the urine and the feces.

WHITE BLOOD CELLS (LEUKOCYTES)

The white blood cells or leukocytes differ from the red blood cells in their abundance, appearance, and physiological functions. In the average adult, one c.c. of blood contains about 5,000,000 leukocytes, and these are subdivided into two major groups: **granulocytes** and **agranulocytes** (Fig. 8-7).

Granulocytes

Granulocytes are white blood cells whose cytoplasm is characterized by the presence of large numbers of small granules. There are three kinds of granulocytes, named according to the properties of the histological stains that are used to visualize and charac-

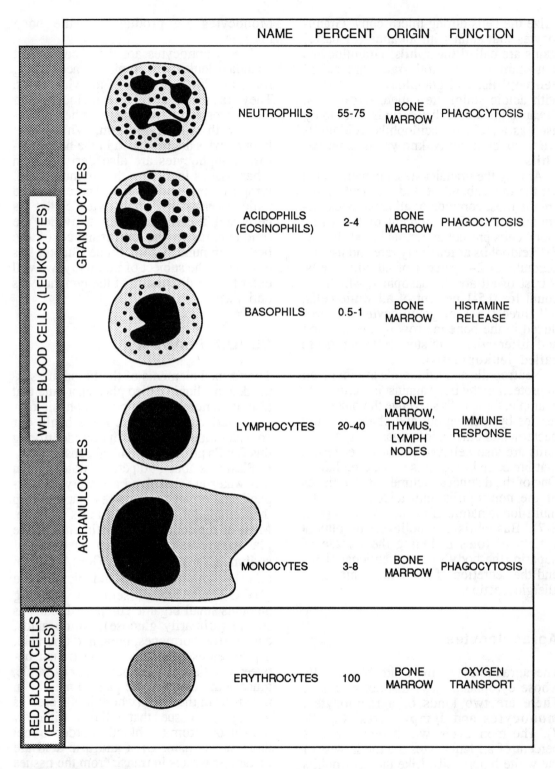

	NAME	PERCENT	ORIGIN	FUNCTION
NEUTROPHILS	55-75	BONE MARROW	PHAGOCYTOSIS
ACIDOPHILS (EOSINOPHILS)	2-4	BONE MARROW	PHAGOCYTOSIS
BASOPHILS	0.5-1	BONE MARROW	HISTAMINE RELEASE
LYMPHOCYTES	20-40	BONE MARROW, THYMUS, LYMPH NODES	IMMUNE RESPONSE
MONOCYTES	3-8	BONE MARROW	PHAGOCYTOSIS
ERYTHROCYTES	100	BONE MARROW	OXYGEN TRANSPORT

WHITE BLOOD CELLS (LEUKOCYTES) — GRANULOCYTES / AGRANULOCYTES

RED BLOOD CELLS (ERYTHROCYTES)

Figure 8-7
Types, relative abundance, origin, and functions of the various kinds of blood cells (leukocytes and erythrocytes).

127

terize the cells during microscopy. Granulocytes that stain with alkaline (or basic) stains are called **basophils**. Granulocytes that stain with neutral stains are called **neutrophils**, and granulocytes that stain with acidic stains are called **acidophils**. Because the acidic dye *eosin* is commonly used as a stain for acidophils, acidophils are also commonly known as **eosinophils**.

Among the granulocytes, the neutrophils are the most abundant. Indeed, neutrophils are the most common of all leukocytes, accounting for 55-75 percent of the white blood cells present in normal blood (Fig. 8-7). Acidophils ar relatively rare and usually account for 2-4 percent of all white cells. Rarest of all are the basophils, which account for 0.5-1 percent of all white cells. All three types of granulocytes are produced in the bone marrow by the division and differentiation of stem cells (a process called **leukopoiesis**).

Neutrophils and acidophils are involved in protecting the body against infection. For example, these cells scavenge the blood and tissues by seeking out and phagocytosing bacteria and other microorganisms. Acidophils are also believed to release enzymes that break up blood clots in the circulation. One of the distinct structural characteristics of the neutrophils and acidophils is the multi-lobed nature of the cell nucleus (Fig. 8-7). Basophils are believed to play a variety of roles including the release of heparin (the body's natural anticoagulant) and the secretion of histamines during an allergic reaction.

Agranulocytes

The agranulocytes are white blood cells whose cytoplasm lacks distinct granules. There are two kinds of agranulocytes: **monocytes** and **lymphocytes** (Fig. 8-7). The monocytes, which represent 3-8 percent of all leukocytes are the largest of the white blood cells. Like the neutrophils and acidophils, the monocytes act as scavengers of foreign organisms and foreign particles by carrying out phagocytosis.

Monocytes are produced in the bone marrow.

The lymphocytes are the second most abundant leukocytes, typically accounting for about 20-40 percent of all white cells. They are smaller than the granulocytes and monocytes and have a large, spherical nucleus with scant cytoplasm. While some lymphocytes are produced in the bone marrow, lymphocytes are also derived from other tissues including the thymus gland, lymph nodes, and spleen. There are two major classes of lymphocytes called **B-cells** and **T-cells** and their activities are fundamental to the performance of the body's immune system. Since Chapter 9 is devoted to the subject of immunity, we will defer further discussion of the lymphocytes until then.

BLOOD PLASMA

The formed elements of the blood are suspended in a liquid called plasma, a fluid that also has a number of important physiological functions of its own. Let's begin our consideration of the plasma by identifying this fluid's principal chemical constituents.

Plasma is about 90 percent water, and in this water are dissolved a variety of inorganic and organic substances that make up the remainder of the plasma's volume. Most of the inorganic constituents of the plasma are salt ions, such as those of sodium (Na^+), potassium (K^+), calcium (Ca^{++}), chlorine (Cl^-), and bicarbonate (HCO_3^-). Also present are varying amounts small organic compounds such as sugars (primarily glucose), amino acids, cholesterol, hormones, urea, and uric acid. These represent substances in transit between one tissue and another. For example, glucose dissolved in the plasma may be in transit from the liver (where it is stored as glycogen) to tissues that will then withdraw the glucose from the blood in order to fuel their metabolism. Urea and uric acid are metabolic wastes in transit from the tissues in which they were produced to the organs that will excrete them from the body (e.g., the kidneys). The hormones of the plasma are chemical messages released into the blood by one tissue (i.e., a gland) and carried in the plasma to "target" tissues where the hormone causes a modification in the behavior of the target tissue.

The largest of the organic substances dissolved in the plasma are proteins, including some that are also hormones. The **plasma proteins** usually account for about 8 percent of the plasma's volume and are of two types: **albumins** and **globulins**. The albumins, which represent about half of the plasma proteins, are highly soluble proteins that are responsible for the plasma's *colloid osmotic pressure* (see below). The globulins are chemically much more diverse than the albumins and also play a greater variety of physiological roles. Three classes of globulins are identified and are termed **alpha, beta,** and **gamma globulins**. The alpha and beta globulins include proteins that bind and shuttle metal ions (e.g., iron [Fe+++] and copper [Cu++]), lipids, vitamins, and other small molecules from one tissue to another; these are often referred to as the *transport proteins*. The gamma globulins are secreted into the blood by certain lymphocytes and act to protect you against infection; these proteins are also known as **antibodies** or **immunoglobulins**.

Colloid Osmotic Pressure

The exchange of nutrients and wastes between the bloodstream and the tissue fluid that bathes the cells of the body occurs at the level of the **capillary beds**. The walls of capillaries are extremely thin and are permeable to most molecules having molecular weights less than about 10,000. This implies that the plasma proteins (nearly all of which have a molecular weight greater than 10,000) cannot cross the capillary wall and enter the surrounding tissue space. Needless to say, the blood cells are also far too large to permeate the capillary walls.

Whenever a fluid rich in dissolved protein (e.g., the plasma) is separated from a fluid that contains little or no protein (e.g.,

the tissue fluid) by a barrier (e.g., the capillary wall) that is permeable to small molecules, an **osmotic pressure** gradient is created across the barrier. This pressure acts to draw water and small molecules through the barrier from the fluid that contains little or no protein into the fluid that is rich in protein. In the case of the capillary beds, the pressure gradient is created by the plasma proteins (mainly albumin) and acts to draw tissue fluid into the blood flowing through the capillaries. The pressure is called **colloid osmotic pressure** (a "colloid" is any substance that will not cross a biological membrane and usually infers that the substance in question is protein).

Colloid osmotic pressure is not the only pressure at play across the walls of capillaries. There is also (1) *blood pressure* (the hydrostatic pressure exerted against the walls of the blood vessels as the blood is driven through the vessel by the actions of the heart), (2) *tissue fluid osmotic pressure* (i.e. the osmotic pressure created by the small amount of protein and other large molecules present in tissue fluid), and (3) *tissue fluid hydrostatic pressure* (i.e., the pressure created in the tissue spaces by the tissue fluid).

The effects of all of these pressures are illustrated in Figure 8-8. For simplicity we may ignore the hydrostatic and osmotic pressures of the tissue fluid, since their values are so small. Instead, we will focus on the blood pressure and the colloid osmotic pressure created by the plasma proteins. Recalling that blood pressure is a pressure pushing *outward* against the capillary walls and that colloid osmotic pressure is a pressure pulling *inward* (into the capillaries) from the tissue space, it is clear that these two pressures oppose one another. In the capillaries at the arterial end of a capillary bed (i.e., the end that is closest to the arteriole feeding blood into the capillaries), the blood pressure is much higher than the colloid osmotic pressure. Therefore, there is a *net* pressure directed across the capillary wall into the tissue space. This pressure difference causes water and small molecules to pass from the bloodstream into the tissue space.

> For example, a typical blood pressure value at the arterial end of a capillary bed is 40 mm Hg, whereas the colloid osmotic pressure rarely exceeds 30 mm Hg; this creates a pressure difference of +10 mm Hg (i.e., 40 minus 30).

By way of contrast, at the venous end of a capillary bed (i.e., the end that is closest to the venule that drains the capillary bed), the blood pressure is considerably lower. The drop in blood pressure is not accompanied by a fall in colloid osmotic pressure because the plasma protein content of the blood remains the same. As a result, the colloid osmotic pressure *exceeds* the blood pressure, thereby creating a net pressure that is directed across the capillary wall from the tissue space into the blood. This pressure difference causes water and small molecules to pass from the tissue space into the bloodstream.

> For example, the blood pressure drops to about 20 mm Hg at the venous end of a capillary bed, whereas the colloid osmotic pressure remains at about 30 mm Hg; this creates a pressure difference of −10 mm Hg (i.e., 20 minus 30).

It should be apparent from the preceding discussion that the tissue fluid bathing the body's cells undergoes continuous turnover. The tissue space receives a con-

tinuous infusion of new materials at the arterial end of a capillary bed, where blood pressure dominates colloid osmotic pressure. At the venous end of the capillary bed, where colloid osmotic pressure dominates blood pressure, tissue fluid is drawn back into the bloodstream. Along with the reabsorbed fluid go the tissue secretions and cellular wastes. As a result, the composition of the fluid bathing the body's cells remains fairly constant. Nutrients that are taken up by the cells are replaced with fresh nutrients from the arterial blood. Secretions and wastes emptied into the fluid by the tissue cells do not accumulate since they are taken up in the venous blood. The relative constancy of the tissue fluid that results is known as **homeostasis**.

BLOOD COAGULATION

We will finish this discussion of the blood by considering one of the blood's most important physiological properties–the capacity to **coagulate**, that is, to undergo a transformation from fluid to solid.

It would be difficult to exaggerate the value of the blood's capacity to coagulate, because it is coagulation that prevents massive and fatal losses of blood from the body following injury. By coagulating and forming a temporary plug that seals a wound, the clot or **thrombus** serves to minimize the loss of blood from the body.

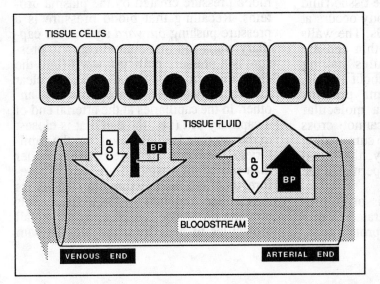

Figure 8-8
The passage of nutrients into the tissue fluid and the passage of cell wastes into the bloodstream are based on an interplay between **hydrostatic pressure** (i.e., blood pressure, **BP**) and **colloid osmotic pressure** (COP). BP acts to drive fluid out of the blood, whereas COP acts to draw fluid into the blood. At the arterial end of a capillary bed, BP dominates COP (large arrow points into the tissue fluid), whereas at the venous end of a capillary bed, COP dominates BP (large arrow points into the bloodstream).

Of course, it is also important that blood does not coagulate while it circulates in the body. Whereas small internal clots do not pose a serious health threat and are quickly removed, a large internal thrombus could obstruct a blood vessel, thereby causing serious harm to the tissues normally fed by that vessel. The body's natural anticoagulant, a substance called **heparin** that is continuously secreted into the blood by the liver, acts to prevent the formation of large internal clots. Purified heparin can also be used to prevent donated blood (or blood drawn from the body for a blood test) from coagulating. Since purified heparin is rather expensive, other chemicals that have been found to possess anticoagulant properties are more commonly used. Among these anticoagulants are citrates, oxalates, and EDTA (ethylenediaminetetraacetic acid). The actions of heparin and other anticoagulants are described later.

Platelets (Thrombocytes)

Tiny formed elements known as **platelets** play an important role in the coagulation process. In a normal person, the platelets number about 250,000,000/c.c. of blood and are released into the circulation from bone marrow. In humans (and other mammals), platelets are not cells; rather they are small fragments of cells that are shed from the surfaces of **megakaryocytes** in the bone marrow (Fig. 8-9). The platelets are rich in secretory granules whose contents help to trigger the coagulation process. In those animals in which platelets are whole cells (i.e., not just fragments of cells), the platelets are more appropriately called **thrombocytes**.

When blood circulates through uninjured tissue, the platelets are repelled by one another and are also repelled by the lining or **endothelium** of blood vessels. However, when there is tissue injury, the damaged sub-endothelial cells release a protein called **von Willebrand factor** that coats the injured surfaces. Platelets stick to this coating, break open, and release their secretory granules.

The secretory granules of platelets contain adenosine diphosphate, thrombaxane A2, and serotonin. Thrombaxane A2 and serotonin are *vasoconstricters* that act to reduce the flow of blood into the injured site. As more and more platelets adhere to the injured tissue a **platelet plug** is formed.

Formation of the Fibrin Network

The blood clot that serves to seal a wound consists of more than just a platelet plug. Also formed is a network of protein fibers, called **fibrin threads**, that support the platelet plug (acting much like scaffolding). The network of fibrin threads also acts to trap red and white blood cells, thereby increasing the size of the clot (Fig. 8-10). The formation of the fibrin thread network is an essential part of the coagulation process.

The fibrin threads of a blood clot are formed from a plasma protein called **fibrinogen** that is manufactured in the liver and continuously secreted into the bloodstream. The chemical conversion of fibrinogen to fibrin occurs in two ways known as the **intrinsic** and **extrinsic pathways** (Fig. 8-11).

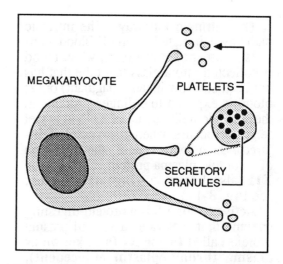

Figure 8-9
Platelets are produced in the bone marrow by the fragmentation of marrow cells called megakaryocytes.

131

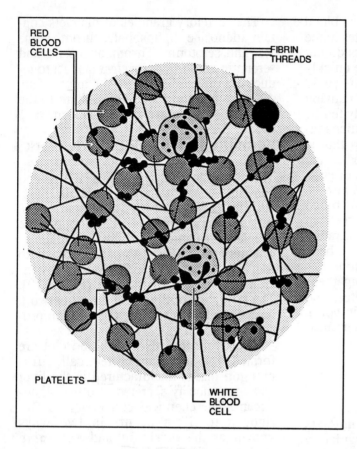

RED
BLOOD
CELLS

FIBRIN
THREADS

PLATELETS

WHITE
BLOOD
CELL

Figure 8-10
A blood clot. A blood clot is formed when **fibrinogen** dissolved in the plasma is converted into insoluble **fibrin threads**. These threads form a meshwork that entraps blood cells (and platelets), thereby forming a plug that seals the wound.

The Intrinsic Pathway. The intrinsic pathway is initiated when (1) blood contacts a foreign surface (e.g., when blood is collected into a glass test tube that has not been coated with anticoagulant) or (2) blood is exposed to collagen proteins in the subendothelial tissue of a blood vessel (e.g., as occurs when the vessel is injured). Such contact causes the activation of a soluble plasma protein called **factor XII** (also known as **Hageman factor**; see Fig. 8-11).

Active factor XII is a protein-digesting enzyme that activates a second plasma protein called **factor XI** (also known as **plasma thromboplastin antecedent**). Activated factor XI then activates a third plasma protein called **factor IX** (or **Christmas factor**). Active factor IX, together with a fourth plasma protein called **factor VIII** (or **antihemophilic factor**) and calcium ions (**factor IV**), then activates a fifth plasma protein called **factor X** (or **Stuart-Prower factor**).

Active factor X, together with a sixth plasma protein called **factor V** (or **proaccelerin**) and additional calcium ions, then converts **prothrombin** (also known as **factor II**) into **thrombin**. Thrombin is a protein-digesting enzyme that converts **fibrinogen (factor I)** into **fibrin monomers**. In a final step, the fibrin monomers are caused to *polymerize* (i.e., to combine with one-another into long chains), thereby forming the fibrin threads that are the basis of the fibrin network. Polymerization of the fibrin monomers is catalyzed by a final plasma clotting factor called **factor XIII** (or **fibrin stabilizing factor**).

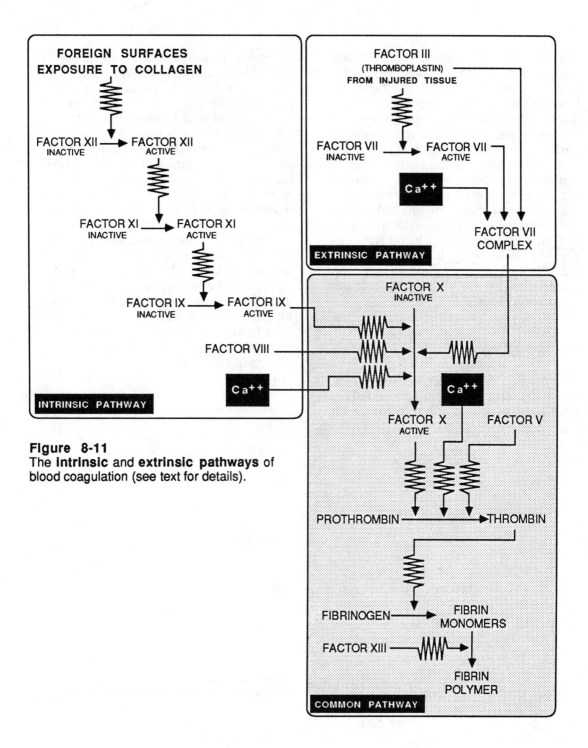

Figure 8-11
The **intrinsic** and **extrinsic pathways** of blood coagulation (see text for details).

The Extrinsic Pathway. The extrinsic pathway is initiated when damaged tissue cells release **thromboplastin (factor III)** into the injury site. Although most tissues contain thromboplastin, thrombo-plastin is especially abundant in skin and in the brain. Injury in these tissues is, therefore, quickly followed by coagulation. Thromboplastin activates **factor VII** (also called **proconvertin**), and then combines with active factor VII and calcium ions to form a large molecular complex that serves to activate factor X. From this point on, the sequence of reactions is identical to that described in the preceding paragraph (i.e., the **common pathway** depicted in Fig. 8-11).

As noted earlier in the chapter, the coagulation of blood may be prevented using anticoagulants such as heparin (the body's natural anticoagulant) and certain salts, such as citrates, oxalates, and EDTA. Heparin prevents blood from coagulating by indirectly inactivating thrombin. Citrates, oxalates, and EDTA stop coagulation by combining with calcium ions (a process called *chelation*).

Hemophilia

Hemophilia is an inherited disease in which the afflicted person's blood fails to coagulate properly. The result is an excessive loss of blood from the body even when an injury is minor. There are two major forms of hemophilia called **hemophilia A** (about 85% of all hemophilias) and **hemophilia B** (about 15% of all hemophilias). In people afflicted with hemophilia A, the problem is an inability to produce properly functioning factor VIII (antihemophilic factor). In hemophilia B (also called **Christmas disease**), there is an inability to produce proper levels of factor IX (Christmas factor). Both of these diseases are inherited. The diseases are more common in males than in females, because the genes governing the production of factors VIII and IX are carried on the X-chromosomes of cells. Hemophilia and other so-called "sex-linked" diseases are considered further in Chapter 14 which takes up the subject of heredity.

CHAPTER 9

THE IMMUNE SYSTEM

In this chapter, we will consider the organization and actions of the body's **immune system**. At the same time, we will take up several closely related subjects including **blood typing**, **blood transfusion**, and **cancer**.

It might be well to ask at the outset what the subjects of immunity and cancer have in common. Immunity is a state attainable in humans (as well as other higher vertebrates) in which the body is protected against the harmful effects of certain disease-causing (i.e., pathogenic) agents such as parasites, bacteria and viruses. Cancer, on the other hand, is a condition characterized by the abnormal and rapid growth of tissue cells to form tumors that in many instances interfere with normal tissue functions and often lead to death. In recent years, it has become clear that the body's system of defense against viruses, bacteria, and other infectious agents (i.e., the immune system) also serves as the principal line of defense against cancer. Moreover, it appears that some cancers are intimately associated with viral infections; that is, certain viruses are **oncogenic** (cancer-creating). Later in the chapter, when we explore the causes of and defenses against cancer, other relationships between the immune system and cancer will become apparent.

The body employs two major systems of defense against infection; these are called **fixed** and **adaptive** defense systems. A fixed system of defense is one that protects the body against infection in a non-specific way. For example, the skin serves as a non-specific barrier against the penetration of different toxic chemicals and most microorganisms. Internally, the blood and other body fluids contain several classes of leukocytes (e.g., the monocytes and neutrophils; see Chapter 8) that serve to scavenge foreign particles (and cancer cells) by phagocytosis. Although the action of these scavengers is not specific (i.e., they may phagocytose any of a variety of foreign agents of disease), their interaction with products of the immune system (see below) increases their specificity and efficiency.

The body's immune system is an adaptive system of defense and is highly specific. Through what is called an **immune response**, agents of infection are recognized and destroyed by either the direct or indirect actions of white blood cells called lymphocytes. Additionally, the immune system exhibits long-term memory of its encounters with agents of infection. This **immunologic memory** permits more rapid response to a second, third, or subsequent attack by a given agent.

DUALITY OF THE IMMUNE SYSTEM

There are two ways in which the immune system responds to infection. In the **cel-**

lular immune response, there is a direct interaction between lymphocytes and invading bacteria, or virus-infected cells. In contrast, the **humoral immune response** acts principally through the secretion of proteins called **antibodies** or **immunoglobulins** into the blood plasma, lymph, and tissue fluid. As we shall see, these antibodies combine with foreign substances called **antigens** that are carried in the surface of or are released by the pathogen. The combination of antibody with antigen then initiates a response leading to the elimination of the antigen and, more importantly, its source.

T-Lymphocytes and B-Lymphocytes

The cellular immune response is mediated by a class of lymphocytes called **T-lymphocytes** or **T-cells**. These cells are derived from stem cells in the blood-producing (hemopoietic) tissues of the embryo (i.e., the liver and bone marrow). From these tissues, the T-lymphocytes migrate to and colonize the **thymus gland** (hence, "T" implies "thymus-derived"). Many T-lymphocytes migrate from the thymus to other lymphoid tissues such as the spleen, tonsils, adenoids, and lymph nodes colonizing these tissues.

Humoral immunity is mediated by **B-lymphocytes** or **B-cells**. In birds, B-lymphocytes are derived from the *Bursa of Fabricius*, an outpocketing of the digestive tract that is colonized by hemopoietic stem cells early in embryonic development (thus, "B" implies "bursa-derived"). In mammals, which have no bursa, the origin of B-lymphocytes remains uncertain, although many B-cells are undoubtedly derived from the bone marrow itself.

The activation of T- and B-lymphocytes by the presence of a foreign agent in the body leads to the production of families of cells that combat the infection. The T-lymphocytes give rise to **cytotoxic, helper**, and **suppressor** cells, whereas the B-lymphocytes give rise to antibody-secreting **plasma cells** (Fig. 9-1). Both T- and B-

lymphocytes also give rise to **memory cells**—cells that are responsible for *immunologic memory*.

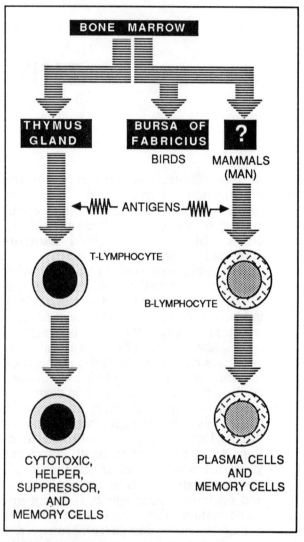

Figure 9-1
Origin of B and T cells (see text for details).

ANTIGENS, ANTIBODIES, AND T-CELL RECEPTORS

An **antigen** is a molecule (or part of a molecule) that acts to stimulate the body's immune system and trigger an immune response. In nearly every instance, the antigen is "foreign" to the person whose immune system is stimulated (i.e., it is chemically different from any of the molecules normally present in the tissues of the person). Although antigens may have diverse chemical character, many are proteins, carbohydrates, and nucleic acids.

Antigens are "recognized" by two families of proteins peculiar to the immune system. The best understood of these proteins are the **antibodies**. Antibodies are found in the plasma membranes of B-cells, where the antibody molecules project from the cell surface and act as antigen receptors. Antibodies are also secreted by B-lymphocytes during an immune response. Because they are produced in large quantities during infection, antibodies are readily isolated for study, and much is now known about their structure and action. Less well understood are the **T-lymphocyte receptor proteins**, proteins that project from the plasma membranes of T-cells. These proteins are not secreted by the T-cells, and they are isolated for study with much greater difficulty. T-cell receptors provide T-lymphocytes with the ability to attach directly to foreign cells that have entered the body and also to the body's own cells when the cells have been infected by viruses.

Antibodies

The antibodies (also called **immunoglobulins**) that circulate in the bloodstream make up part of the "gamma globulin" fraction of blood plasma (indeed, antibodies account for about 20% of the protein present in blood plasma). The production of antibodies is stimulated by antigens that are present in or released from pathogens that have entered the body; typically, the antigens are constituents of bacterial membranes or the coats of viruses and have been recognized by the immune system as being foreign or alien to the body.

The antibodies produced during an immune response have the capacity to bind to the type of antigen that triggered the response. The reaction between the antibody and antigen is highly specific; each type of antibody reacts with a particular antigen and no other. The human body is capable of synthesizing millions of different kinds of antibody molecules, each antibody capable of reacting with a different antigen.

Although millions of different antibody molecules can be produced by the human immune system, all can be assigned to one of five antibody classes on the basis of chemical and functional similarities. These classes are called *IgA, IgD, IgE, IgG,* and *IgM* ("Ig" stands for "immunoglobulin," a term used interchangeably with the term "antibody;" "A," "D," "E," "G," and "M" stand for alpha, delta, epsilon, gamma, and mu [letters of the Greek alphabet]). The most abundant antibodies are those that belong to the IgG class (about 80% of all antibodies), and the organization of the typical IgG molecule is depicted in Figure 9-2. Each antibody is composed of four polypeptide chains having two different sizes; there is a pair of identical high molecular weight chains called **heavy chains**

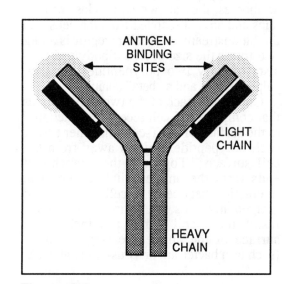

Figure 9-2
Organization of an antibody molecule of the IgG class. The molecule contains 4 polypeptide chains: two heavy (i.e., large) chains and two light (i.e., small) chains. The interaction between antibody and antigen occurs at binding sites shared by the ends of heavy and light chains.

137

and a pair of identical low molecular weight chains called **light chains**. (The heavy chains contain about 450 amino acids, whereas the light chains consist of about 214 amino acids.) Each light chain is linked to a heavy chain, and the two light chain/heavy chain pairs are linked together.

The chemical reaction that takes place between antigen and antibody occurs at the ends of the paired heavy and light chains of the antibody molecule (see Fig. 9-2). It is in this region that the numbers, kinds, and sequences of amino acids that comprise the heavy and light polypeptide chains differ from one antibody molecule to another. Much of the remaining chemical structure of the antibody molecule is the same in all members of an antibody class.

T-Cell Receptors

Like the plasma membranes of B-cells, T-cell surfaces also contain antigen-binding receptors. Although **T-cell receptors** have many properties in common with antibodies, they are a distinct class of proteins and are not secreted from the cells into the bloodstream. A T-cell receptor is about two-thirds the size of an antibody and consists of only two polypeptides (i.e., an "alpha" chain and a "beta" chain).

One end of each of the two polypeptides that comprise a T-cell receptor is anchored to the surface of the T-cell; the other end of each polypeptide projects away from the cell surface. Together, the unanchored ends form the antigen-binding site. The interaction between a T-cell receptor and antigen involves the entire T-cell. As a result, the T-cell is able to attach to the surface of the antigen-bearing pathogen (such as a bacterium or virus-infected cell).

CLONAL SELECTION THEORY

The **clonal selection theory** explains the mechanism by which the body's immune system responds to the appearance of antigens present in or released from an agent of infection. The immune system in- cludes many millions of different **clones** of B- and T-lymphocytes, each clone consisting of cells that are capable of responding to the same antigen. Thus, the body's immune system is continuously "on alert" in anticipation of the appearance of any of more than a million different antigens.

The immune response is triggered when antigens of the infecting agent become bound to immunoglobulin molecules in the plasma membranes of B-cells or to T-cell receptors. Usually, a given antigen can be bound by the surface receptors of a number of different clones, but this is a tiny percentage of the total number of clones present. Thus, the clones that respond to the presence of the antigen are, in effect, "selected" by the antigen, and this is what is meant by "clonal selection."

Binding of antigen to the surface receptors of the cells of a particular B- or T-lymphocyte clone serves to activate these cells causing a succession of rounds of cell division that greatly increases the numbers of cells of that clone that are present in the body (Fig. 9-3). In the case of B-cells, some of the newly-produced cells become **plasma cells** that synthesize and secrete antibodies that react with additional antigen. Other cells produced by divisions of the selected clone become **memory cells** that are reserved for a response to a subsequent exposure to the same antigen. In the case of T-cell clone selection, some of the newly-produced T-cell also become memory cells, but most develop into **cytotoxic, helper**, and **suppressor cells** responsible for the cell-mediated immune response.

B-LYMPHOCYTES AND THE IMMUNE RESPONSE

To understand how B-lymphocytes are caused to secrete antibodies, let's consider a case in which a person acquires either a bacterial or viral infection. Antigens present on the surface of (or released by) the pathogen become bound to antibodies in the plasma membranes of one or more of the millions of clones of B-lymphocytes. Binding of the antigen to the surface of the

B-lymphocytes does not by itself cause activation of the clone. Rather, antigens must also be taken up during non-specific phagocytosis of antigen-bearing particles by **macrophages** (i.e., phagocytic cells such as granulocytes and monocytes that act as scavengers in the body's tissues; see Chapter 8). The antigens taken up by the macrophages are degraded and fragments of the antigens containing **antigenic determinants** are then displayed at the cell surface (Fig. 9-4). Macrophages that carry out this process are referred to as **antigen-presenting cells**. The antigenic determinants are recognized by one or more clones of T-cells possessing T-cell receptors for the antigen. T-cells that recognize and are activated by antigen-presenting cells are called **T helper cells**. Activated T helper cells then interact with the B-lymphocytes to which antigen had already been bound. The interaction between T helper cells and B-lymphocytes serves to activate the B-lymphocytes causing the rapid multiplication of members of the clone, thereby yielding large numbers of plasma cells and memory cells (Fig. 9-4). Only the plasma cells produce and secrete antibodies. The memory cells are kept in reserve and will be called on to respond during a second (or subsequent) infection by the same antigen-bearing pathogen.

Figure 9-3
The "clonal selection" theory (see the text for discussion).

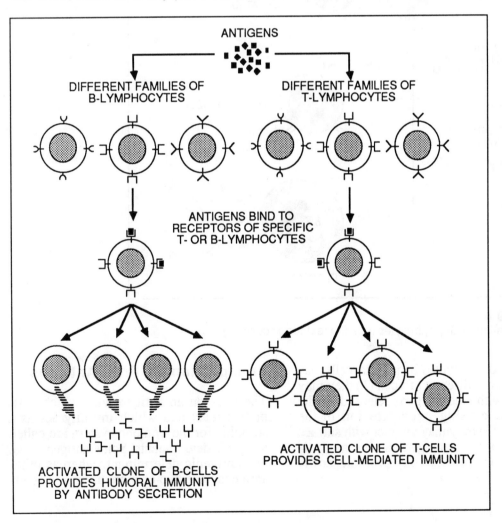

ANTIGENS

DIFFERENT FAMILIES OF B-LYMPHOCYTES

DIFFERENT FAMILIES OF T-LYMPHOCYTES

ANTIGENS BIND TO RECEPTORS OF SPECIFIC T- OR B-LYMPHOCYTES

ACTIVATED CLONE OF B-CELLS PROVIDES HUMORAL IMMUNITY BY ANTIBODY SECRETION

ACTIVATED CLONE OF T-CELLS PROVIDES CELL-MEDIATED IMMUNITY

139

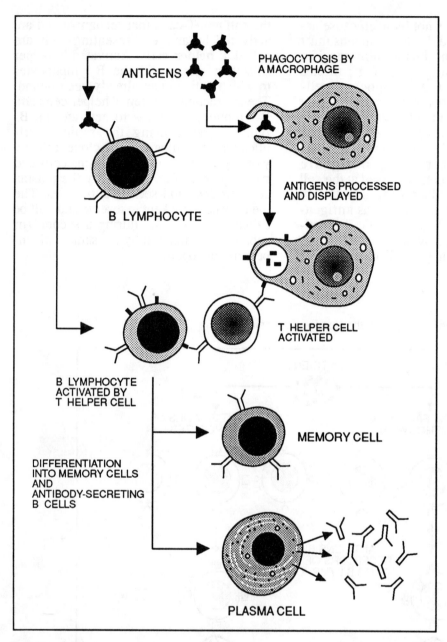

Figure 9-4
The activation of B-lymphocytes (see the text for discussion).

Antibodies secreted by plasma cells may have several different effects: (1) they may interact with free antigens released by the pathogen causing **precipitation** of the antigens; (2) they may interact with surface antigens of the pathogen causing **agglutination**; or (3) they may promote **complement fixation**.

Precipitation of Free Antigens

The sites on an antigen with which antibodies react and which initially act as a stimulus for the immune system are called antigenic determinants. Some antigens have one antigenic determinant, whereas other antigens have two or more antigenic determinants. If one antigenic determinant is present, the antigen is said to be **mon-odeterminant;** if two are present, the antigen is **bideterminant**, and so on. Most antibodies are **bivalent**, meaning that they can simultaneously combine with up to two antigenic determinants. The products formed by interaction of antibody and antigen depends upon the number of antigenic determinants that are present. Two monodeterminant antigens can be cross-linked by a single antibody molecule, but the product is not usually insoluble. However, if two antigenic determinants are present, cross-linking by the antibody can produce chains of antigens that are insoluble and form precipitates. When three (or more) antigenic determinants are present on the antigens, the reaction with antibodies produces cross-linked networks that are insoluble and form precipitates. This process is illustrated in Figure 9-5.

Agglutination

Antibodies that interact with antigens present in the surfaces of invading microorganisms cause **agglutination** (Fig. 9-6). During agglutination the microorganisms become cross-linked to form small masses. These masses are phagocytosed by macrophages (i.e., monocytes), which kill and digest the microorganisms. As illustrated in Figure 9-6, the plasma membranes of macrophages possess receptors that bind the paired ends of the antibody heavy chains (i.e., they bind to the ends of the heavy chains that are opposite to the antigen binding sites).

Although the mechanism is not fully understood, foreign cells to which antibodies have attached can also be destroyed by **K (or killer) cells.** Killer cells bind the foreign cells to which antibody has already attached by interacting with the same regions of the antibody molecules as do the macrophages. However, killer cells do not internalize the foreign cells. Instead, it is thought that there is the transfer of toxic substances from the K cell to the pathogen.

Complement Fixation

Complement fixation system is yet another mechanism by which antibodies defend the body against invasion by pathogens. **Complement** is a family of more than a dozen proteins that circulate in

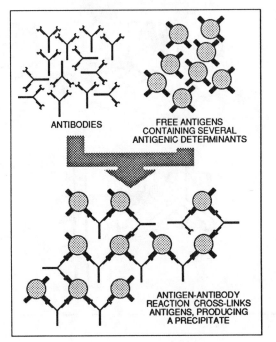

ANTIBODIES

FREE ANTIGENS
CONTAINING SEVERAL
ANTIGENIC DETERMINANTS

ANTIGEN-ANTIBODY
REACTION CROSS-LINKS
ANTIGENS, PRODUCING
A PRECIPITATE

Figure 9-5
Antibodies can cause the precipitation of free antigens. Illustrated here is a reaction between antibodies and free antigens that contain several **antigenic determinants**. The antigens are *cross-linked* by the antibodies, thereby producing a precipitate.

141

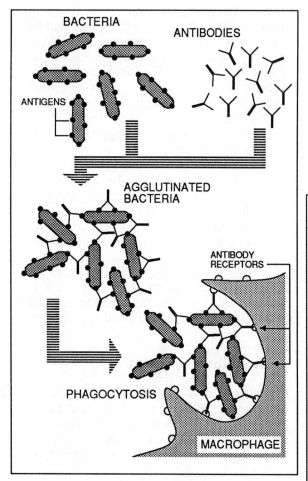

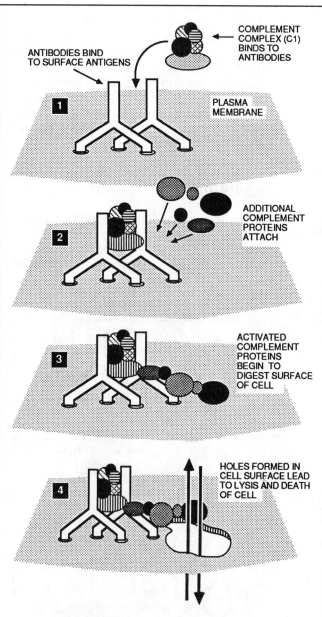

Figure 9-6
In some instances, antibodies react with antigens in the surfaces of invading bacteria. The bacteria may be cross-linked by these reactions, thereby producing an *agglutinated* mass that is subsequently phagocytosed by the body's macrophages.

BACTERIA

ANTIBODIES

ANTIGENS

AGGLUTINATED BACTERIA

ANTIBODY RECEPTORS

PHAGOCYTOSIS

MACROPHAGE

ANTIBODIES BIND TO SURFACE ANTIGENS

COMPLEMENT COMPLEX (C1) BINDS TO ANTIBODIES

PLASMA MEMBRANE

ADDITIONAL COMPLEMENT PROTEINS ATTACH

ACTIVATED COMPLEMENT PROTEINS BEGIN TO DIGEST SURFACE OF CELL

HOLES FORMED IN CELL SURFACE LEAD TO LYSIS AND DEATH OF CELL

Figure 9-7
During **complement fixation**, antibodies that bind to the surfaces of foreign cells trigger the progressive assembly of plasma complement proteins at the cell's surface. The assembled protein complex then digests the foreign cell's surface, thereby causing the cell's lysis.

the blood (unlike the immunoglobulins, complement proteins are not synthesized in response to the appearance of an antigen). The binding of antibodies to a cluster of antigenic determinants in the surfaces of bacteria triggers a cascade of reactions in which the complement proteins (many of which are digestive enzymes in an inactive state) are sequentially bound to the pathogen and are activated (Fig. 9-7).

The reaction cascade is initiated by the binding of a small complex of the complement proteins to antibodies that are bound to the bacterial antigens. In the ensuing reactions, additional complement proteins are bound and activated, eventually forming a **lytic complex** that creates an open channel through the bacterial surface. By disorganizing the bacterium's plasma membrane and allowing water to enter the cell by osmosis, complement causes the cells to swell and lyse (i.e. burst), thereby killing the cell. Complement fixation by antibody-coated bacteria and the lysis of the invading cells that follows is the most common defense mechanism attributable to B-cell-secreted antibodies.

Immunologic Memory

Figure 9-8 describes the relationship between elapsing time and the appearance of antibodies in response to a given antigen. Following a short lag period, antibodies begin to appear in the blood, rising to and maintaining a plateau level for some time before falling again. This characteristic response curve is called a **primary immune response.** So long as the antibody content of the blood remains at its plateau level, a condition of **active immunity** exists. The response to a second exposure to the same antigen–the **secondary immune response**–is more dramatic. The lag period is shorter, the response is more intense (greater quantities of antibody are produced), and the elevated antibody level is maintained for a longer period of time. The difference between the two responses indicates that the body has "remembered" its earlier encounter with the antigen.

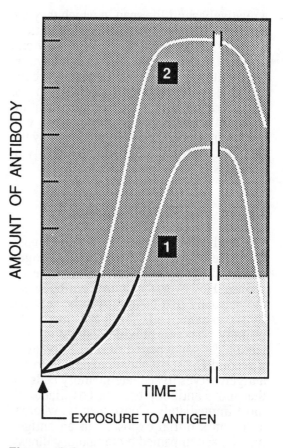

Figure 9-8
Primary immune response occurs shortly after exposure to antigen and is characterized by secretion of antibodies into the blood (curve [1]). The level of circulating antibodies rises and soon exceeds the amount necessary to react with all antigen present (grey zone near baseline). Excess antibody is retained for some time but eventually declines. A second exposure to the same antigen is followed by more rapid production of antibodies, and the amount of antibody produced exceeds that which followed the first exposure (curve [2]). The more rapid **secondary immune response** is attributed to memory cells produced during the primary immune response.

Immunologic memory may be explained in the following way. The initial exposure to antigen causes differentiation of B-lymphocytes into memory cells as well as plasma cells. Whereas the plasma cells have

143

a relatively short life span in which they are actively engaged in antibody secretion, memory cells do not secrete antibody and continue to circulate in the blood and lymph for months or years. These memory cells are able to respond more quickly to the reappearance of the same antigen than undifferentiated B-lymphocytes. As we will see, memory cells are also produced by the multiplication and differentiation of T-lymphocytes.

Active vs. Passive Immunity

The antibodies produced during an immune response serve to destroy the pathogen whose antigens initially triggered the response. However, the unused antibody and the antibody that continues to be secreted by the plasma cells produced during the primary response provide continuing protection for some period of time. Moreover, a second (or subsequent) exposure to the antigen serves to activate memory cells, so that more antibody becomes available almost immediately. Consequently, whereas a person's initial exposure to an antigen may be accompanied by temporary illness, subsequent exposures to the same antigen are handled by the immune system with relative ease. This is what is meant by **active immunity**—the ability to easily fend off the harmful effects of a pathogen to which you have previously been exposed and to which you developed a primary response. For example, once you have had chicken-pox, subsequent exposure to the chicken-pox virus is rarely followed by any symptoms of the illness. For some pathogens (e.g., chicken-pox virus and measles virus), active immunity lasts a lifetime.

Active immunity is to be distinguished from **passive immunity**. Passive immunity is conferred to people who have been exposed to extremely dangerous pathogens (e.g., snake venom, tetanus bacteria, etc.) and whose immune systems may not react quickly enough to the pathogen to prevent serious harm. The immunity is conferred by injecting **antiserum** containing anti-

bodies against the pathogen into the bloodstream of the infected person (e.g., receiving a "tetanus shot"). The antibody-rich antiserum is usually obtained by processing blood from an animal (e.g., horse, dog, cow, etc.) that has deliberately been exposed to the antigen and which developed a primary immune response. The antibodies injected into the bloodstream of the infected individual provide an immediate (although temporary) defense against the pathogen.

Autoimmune Diseases

One's immune system normally produces antibodies against foreign proteins but not against the native proteins of the body. That is, the immune system is able to distinguish between "self" and "non-self." Yet, one's own proteins may be regarded as antigens by the immune system of another person (or another animal). Thus, each person's tissues possess proteins (and other chemical substances) that are potential antigens. The ability to distinguish self from non-self develops very early in life.

In rare cases, individuals begin to produce antibodies against their own antigens. These antibodies are called **autoantibodies** and the diseases resulting from their presence are called **autoimmune diseases**. Among the autoimmune diseases are **paroxysmal cold hemoglobinuria** (antibodies against one's own red blood cells), **myasthenia gravis** (antibodies against one's own muscle cell acetylcholine receptors), and **systemic lupus erythematosus** (antibodies against one's own DNA).

MAJOR HISTOCOMPATIBILITY COMPLEXES

As already noted, B- and T-cells contain receptors that react with specific antigenic determinants. B-cells are activated by interaction with the antigen either in its free form or while it is still a part of the surface of a pathogen. In contrast, the T-cells are activated only when the antigen is displayed

144

on the surface of a cell that carries markers of the person's own identity. These markers are the **major histocompatibility proteins** (or **MHC proteins**).

Different (i.e., unrelated) individuals have different sets of MHC proteins; thus, like the antibodies and T-cell receptors, the MHC proteins are incredibly diverse. However, unlike antibodies and T-cell receptors, which differ among the millions of different clones of cells of an individual, the MHC proteins differ only among individuals; that is, all cells of a single individual carry the same MHC proteins.

MHC proteins are divided into two major classes: Class I MHC antigens and Class II MHC antigens. Class I antigens are found in the surfaces of nearly all cells, whereas class II antigens are limited to a few types of cells that play a role in the immune response. For example, class II antigens are found in most B-cells, some T-cells, and some antigen-presenting macrophages.

T-LYMPHOCYTES AND THE IMMUNE RESPONSE

T-cells do not interact with free antigens or with antigenic sites in the surfaces of foreign microorganisms. Instead, T-cells respond only to cells carrying both a self MHC antigen and an antigenic determinant from a foreign source (i.e., from bacteria, viruses, etc.). Thus, two stimuli are needed to trigger the proliferation of the required T-cell clones. **Cytotoxic T-cells** respond to the combination of foreign antigen and class I MHC antigens, whereas **helper T-cells** respond to foreign antigen and class II MHC antigens. Thus, the activities of the T-cells are directed toward the body's own cells and not to free pathogen.

Cytotoxic T-Cells

The actions of cytotoxic T-cells are illustrated by the body's response to viral infection (Fig. 9-9). When a virus attacks a cell, the viral nucleic acid enters the host cell, whereas the proteins of the viral coat remain at the cell surface. Consequently, the infected cell has the proper combination of surface antigens to be recognized by cytotoxic T-cells–namely, class I MHC antigens and viral antigens.

Cytotoxic T-cells attach to newly infected host cells, killing them (presumably through the transfer of **cytotoxins**) before virus replication occurs. A number of host cells are necessarily killed by this process before the virus infection is attenuated. A single cytotoxic T-cell may kill several host cells. Both the viral antigen and the class I MHC antigen are involved in the attachment of the T-cell to the infected host cell.

Helper T-Cells

As noted earlier in the chapter, macrophages and other scavengers phagocytose and degrade foreign antigens and display the antigenic determinants at the cell surface. The combination of antigenic determinant and class II MHC antigens at the surfaces of these cells leads to the attachment and activation of helper T-cells (Fig. 9-10). Helper T-cells can activate B-lymphocytes and other T-cells (e.g., cytotoxic and suppressor T-cells). Certain helper T-cells secrete **lymphokines** or **interleukins**; these are substances that activate macrophages and other lymphocytes. Some lymphokines attract macrophages to the site of infection. Other lymphokines prevent migration of macrophages away from the site of infection. Still other lymphokines stimulate T-cell multiplication. The net effect of lymphokine secretion is the accumulation of macrophages and lymphocytes in the region of an infection and is characterized by the inflammation that typically exists there.

Acquired Immune Deficiency Syndrome (AIDS)

No disease in recent memory has attracted so much public concern and fear as **acquired immune deficiency syndrome**

145

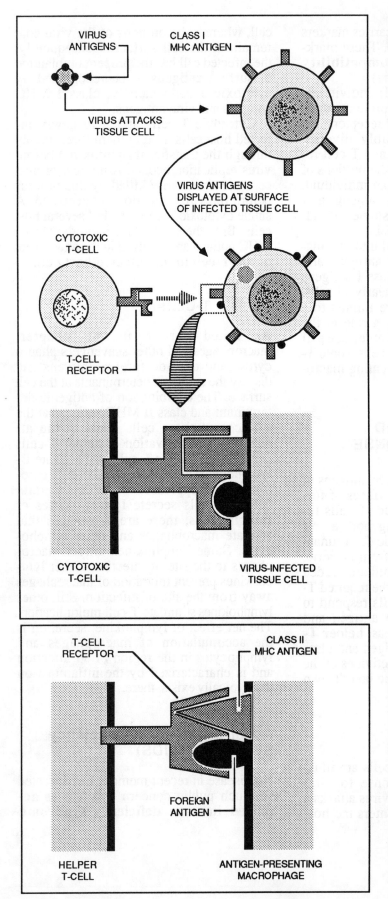

VIRUS ANTIGENS

CLASS I MHC ANTIGEN

VIRUS ATTACKS TISSUE CELL

VIRUS ANTIGENS DISPLAYED AT SURFACE OF INFECTED TISSUE CELL

CYTOTOXIC T-CELL

T-CELL RECEPTOR

CYTOTOXIC T-CELL

VIRUS-INFECTED TISSUE CELL

Figure 9-9
Actions of **cytotoxic T-cells**. Whereas the core nucleic acid of a virus enters the host cell during infection, the viral protein coat is left in the host cell's surface and is displayed along with the cell's class I MHC antigens. The immune system's cytotoxic T-cells recognize this combination of surface markers on the infected cell. Attachment of the cytotoxic T-cell to the infected cell is followed by the transfer of cytotoxins (cell poisons) to the infected cell, killing the cell and, in so doing, preventing the formation of new viruses. Depicted in the lower half of the figure is the interaction between the surface receptor of a cytotoxic T-cell and the foreign antigen/class I MHC antigen of the infected host cell.

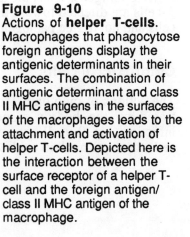

T-CELL RECEPTOR

CLASS II MHC ANTIGEN

FOREIGN ANTIGEN

HELPER T-CELL

ANTIGEN-PRESENTING MACROPHAGE

Figure 9-10
Actions of **helper T-cells**. Macrophages that phagocytose foreign antigens display the antigenic determinants in their surfaces. The combination of antigenic determinant and class II MHC antigens in the surfaces of the macrophages leads to the attachment and activation of helper T-cells. Depicted here is the interaction between the surface receptor of a helper T-cell and the foreign antigen/class II MHC antigen of the macrophage.

146

or **AIDS**. The disease is caused by a virus called **human T-cell lymphotropic virus-III** or **HTLV-III** (HTLV-I and HTLV-II viruses are discussed later in the chapter in connection with cancers of the immune system). HTLV-III critically damages a person's immune system by infecting and eventually killing T-cells. As a result of the progressive destruction of its T-cells, the body is easily ravaged by a number of common infectious agents. In many instances, these infections would have caused little injury were there functional T-cell clones available. Unable to battle infections in the normal manner, victims that develop a "full blown" case of AIDS eventually die. In AIDS patients, the HTLV-III virus has been shown to be present in saliva and semen as well as in the blood. While the virus is believed to be transmitted principally through sexual contact, a number of hemophiliacs have contracted the disease as a result of having received transfusions of infected blood. Moreover, a number of infants have developed AIDS by transplacental transmission from infected mothers.

BLOOD TYPING AND BLOOD TRANSFUSION

The surfaces of red blood cells contain antigens that serve as the basis for blood typing; the distributions of these antigens also dictate when it is safe (or unsafe) to transfuse blood from one person into another. Although there are several blood type antigen families, we will be concerned here with the two most important families known as the **ABO series** and the **Rhesus factor**.

The ABO Series

The ABO series consists of two antigens: **antigen A** and **antigen B**. The presence or absence of one, the other, both, or neither of these antigens dictates each person's blood type, and this is determined genetically (see Fig. 9-11). That is, each of us

has inherited from our parents two genes (one gene from the father and one from the mother) that determine which (if any) antigens of the ABO series our red cells will produce.

There are three different types of ABO genes that can be inherited: an A gene (which dictates that type A antigens will be made by the red cells), a B gene (which dictates that type B antigens will be made by the red cells), and an O gene (this gene does not express itself in the production of any detectable antigen). If a person inherits an A gene from each parent, then that person's **genotype** is said to be AA and his blood type is A. If a person inherits a B gene from each parent, then that person's genotype is said to be BB and his blood type is B. If a person inherits an A gene from one parent and a B gene from the other parent, then that person's genotype is said to be AB and his blood type is AB. It is possible for a person to inherit an A or B gene from one parent but inherit an O gene from the other parent. If the combination is A and O (i.e., genotype AO), then the person's blood type is A. If the combination is B and O (i.e., genotype BO), then the person's blood type is B. Finally, one can inherit an O gene from each parent (i.e., the resulting genotype is OO); such a person has type O blood. Therefore, as seen in Figure 9-11, there are two genotypes that produce type A individuals (i.e., genotypes AA and AO) and two genotypes that produce type B individuals (i.e., genotypes BB and BO). In contrast, there is only one genotype that results in type AB (i.e., genotype AB) and only one genotype that results in type O (i.e., genotype OO).

Proteins that are very similar to the human A and B antigens are found in animal and plant tissues and in the walls of certain bacteria, and this has important physiological consequences. Within a year or two after birth, as these antigens are encountered in food that is eaten (or even in the air that is breathed, because air may contain suspended plant pollen), a child's immune system begins to respond to the "non-self" antigen. For example, a child with type A blood does not respond to A antigens encountered in food, because the A antigens are recognized as "self;" however, B antigens in the food trigger an immune response in which anti-B antibodies are manufactured and secreted into the bloodstream. The production of these antibodies continues through life as the immune system is repeatedly exposed to the B antigens. Consequently, a person with type A antigens in his red cells has anti-B antibodies in his blood plasma. Similarly, a person with type B antigens in his red cells has anti-A antibodies in his blood plasma. A person with type A and type B antigens in his red cells (i.e., blood type AB) will not develop an immune response to either the A or B antigen because both of these are recognized as self. Therefore, a type AB person has neither anti-A nor anti-B antibodies in his blood plasma. Finally, the immune system of a type O person will respond to both the A and B antigens encountered in food and will have both anti-A and anti-B antibodies in his blood plasma (Fig. 9-11).

When the combinations of antigens and antibodies present in the blood of two people are *compatible*, it is possible to successfully *transfuse* blood from one person into the other. In a transfusion, the person receiving the transfused blood is called the **recipient** and the person whose blood is used in the transfusion is called the **donor**. When determining whether or not a particular combination of donor and recipient is compatible, one takes into account the antigens present in the donor's blood cells and the antibodies present in the recipient's blood plasma. It is not necessary to consider the donor's antibodies (or the recipient's antigens) because the blood plasma of the donor is so greatly diluted by the recipient's plasma.

For example, the transfusion of blood from one type A person into another type A person is compatible because the recipient's blood plasma does not contain anti-A antibodies. However, the transfusion of type B blood into a type A recipient would not be compatible because the recipient's blood contains anti-B antibodies. These antibodies would agglutinate the red blood cells in the donated blood creating small masses that can block the smaller blood vessels and interfere with the normal flow of blood in the body. Moreover, the removal of products that result from the breakdown of the agglutinated cells places an excessive strain on the kidneys and other tissues and can result in organ damage. The compatibility/non-compatibility relationships of other combinations of donor and recipient are shown in Figure 9-12.

As you examine Figure 9-12, note that all of the boxes in the AB recipient column contain plus signs (i.e., blood from any of the four ABO blood types can be compatibly transfused into an AB recipient). This is because a type AB person has no anti-A or anti-B antibodies in his blood plasma. It is for this reason that type AB is known as the **universal recipient** blood type. Also note that all of the boxes in the O donor row contain plus signs (i.e., type O blood can be compatibly transfused into any recipient type). This is because the blood cells in type O blood have neither A nor B antigens. For this reason type O is known as the **universal donor** blood type.

Figure 9-11
Antigen and antibody combinations among individuals belonging to each of the four blood types of the ABO series.

BLOOD TYPE	ANTIGENS PRESENT	ANTIBODIES PRESENT	POSSIBLE GENOTYPE	RELATIVE FREQUENCY BLACKS	WHITES
A	A	anti-B	AA AO	27	41
B	B	anti-A	BB BO	20	10
AB	A and B	NONE	AB	7	4
O	neither A nor B	anti-A and anti-B	OO	46	46

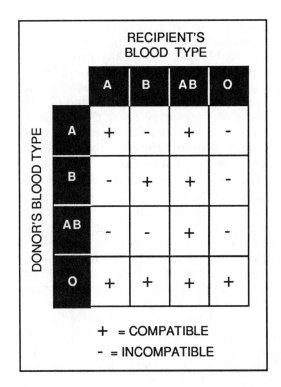

Figure 9-12
Compatible (+) and incompatible (-) ABO series blood transfusions. In an incompatible transfusion, the red blood cells of the donated blood are agglutinated by antibodies in the recipient's plasma.

The Rhesus Factor

Another important red blood cell antigen is the **Rhesus factor** (named after the *Rhesus* monkey in which the antigen was first discovered). Every person is either Rhesus positive (i.e., **Rh+**) or Rhesus negative (i.e., **rh-**), and like the ABO series of blood types, one's Rhesus blood type is determined genetically. Two genes are involved—one that is expressed through the production of the Rhesus antigen (i.e., Rh+) and one that produces no antigen (i.e., rh-). There are, therefore, three possible genotypes (Fig. 9-13): Rh+Rh+ (Rh+ genes inherited from each parent); Rh+rh- (an Rh+ gene inherited from one parent and an rh- gene from the other), and rh-rh- (rh- genes inherited from both parents). So long as at least one Rh+ gene is inherited, the person's blood cells will contain the Rhesus antigen and be classified as Rhesus positive (about 85% of the population). If no Rh+ gene is present, the person is Rhesus negative (about 15% of the population).

A major difference between the ABO series antigens and the Rhesus factor is that a person who is rh- does not automatically develop antibodies against the Rhesus antigen (recall that the plasma of a person who is type A contains anti-B antibodies; the plasma of a person who is type B contains anti-A antibodies; and so on). In order for an rh- person to develop antibodies against the Rhesus antigens, Rh+ blood must have entered his body (as might occur during a transfusion). For example, if a person whose blood type is A/rh- (i.e., his ABO series type is A and he is Rhesus negative) receives a transfusion of A/Rh+ blood, the Rh+ blood cells will not be agglutinated (assuming, of course, that this is the *first* time that he receives a transfusion of Rh+ blood). However, this transfusion **sensitizes** him against Rhesus antigens. That is, the Rhesus antigens in the donated red blood cells trigger an immune response in which **anti-Rhesus antibodies** progressively appear in the blood plasma. As a result, should he receive another transfusion of Rh+ blood, his anti-Rhesus antibodies will agglutinate the donated red cells. For this reason, it is advisable that blood transfusions involve blood types that are both ABO compatible and Rhesus compatible.

BLOOD TYPE	ANTIGENS PRESENT	POSSIBLE GENOTYPE	
Rh+	Rhesus	Rh+Rh+	Rh+rh-
rh-	no Rhesus	rh-rh-	

Figure 9-13.
The Rhesus blood types (see text for details).

Erythroblastosis fetalis

The Rhesus factor takes on special significance during pregnancies in which the mother is rh- and the developing fetus is Rh+. As the fetus' vascular system develops, Rhesus antigen-bearing red cells appear in the fetal circulation. These cells pick up oxygen from the mother's bloodstream as the cells circulate through the capillary beds of the placenta. (The respective bloodstreams of the mother and fetus do not mix, because they are separated by the placental membranes.) The oxygen picked up by the fetal red cells is then carried back to the fetus where it is consumed during metabolism.

If some of the fetal red blood cells (or even small fragments of the cells) pass through the placental membranes separating the fetal circulation from the mother's circulation, the Rhesus antigens will trigger an immune response by the mother. This is because the mother's tissues perceive the fetal Rhesus antigens as "non-self," and so her immune system begins to produce antibodies against Rhesus antigens. Anti-Rhesus antibodies that make their way across the placental membranes from the mother's blood into the fetus' blood will cause agglutination and destruction of the fetus' red blood cells (Fig. 9-14). The fetus becomes anemic, the anemia worsening as more and more of the fetal red cells are lost. The resulting condition is called erythroblastosis fetalis, and if the anemia is severe, it may be fatal to the fetus.

It should be emphasized that erythroblastosis can occur only when the mother is rh- and the fetus is Rh+. If the mother is Rh+, then she perceives any fetal Rhesus antigens that enter her bloodstream as "self" and her immune system does not produce anti-Rhesus antibodies. Of course, the mother will not generate an immune response if both she and the fetus are rh- because there are no Rhesus antigens in the fetus' red cells.

Even when the mother is rh- and the fetus is Rh+, erythroblastosis fetalis rarely occurs during a first such pregnancy.

Among the reasons for this are: (1) fetal Rhesus antigens may not get into the mother's bloodstream at any time during the pregnancy; hence, there would be no opportunity for the mother to become sensitized to the Rhesus antigens; and (2) the Rhesus antigens may enter the mother's bloodstream so late in the pregnancy that the baby is born before a substantial immune response is mounted.

The most likely time at which an rh- mother can be sensitized by the Rh+ antigens present in the blood of the fetus that she is carrying is when she gives birth to the child (i.e., at **parturition**). Birth of the child is accompanied by the loss of some blood on the part of both the mother and the child, and this affords the greatest opportunity for fetal blood cells to enter the mother's tissues. If she is sensitized at that time, then a second (or subsequent) pregnancy in which she again carries an Rh+ fetus presents the greatest risk of erythroblastosis.

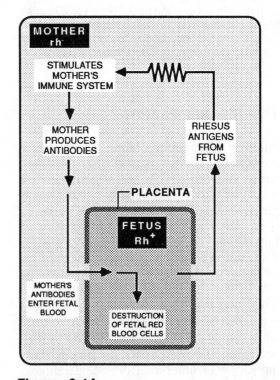

Figure 9-14
Events leading to the ocurrence of Erythroblastosis fetalis (see text for details).

The prospects of erythroblastosis occurring during a second pregnancy in which the mother carries an Rh+ fetus can be diminished greatly by injecting anti-Rhesus antibodies into the mother's blood within 3 days of giving birth to the first Rh+ child. The anti-Rhesus antibodies that are injected into the mother destroy any Rhesus antigens that "leaked" into her bloodstream during parturition, and she is therefore prevented from mounting an immune response to the antigens.

CANCER

Cancer is a disease in which there is excessive and abnormal growth of certain tissue cells. In the healthy tissues of an adult, growth is regulated in such a way that cell multiplication exactly balances cell loss. Consequently, once a person reaches adult age, the sizes and cellular contents of the various organs of the body remain more or less constant. Various chemical, physical, and viral agents can cause the loss of control of cell growth, and as a result, a normal cell may be *transformed* into a cancerous one. The transformed cell undergoes uncontrolled growth and division producing a large clone of transformed cells that is now identified as a **tumor** or **neoplasm**. *Malignant* tumor cells are cancer cells that spread to neighboring tissues–a process that is called **metastasis**. In contrast, *benign* neoplasms consist of cancer cells that do not spread to distant sites and do not pose so great a threat to life. The body's major line of defense against cancer cells is the immune system, which recognizes the changes in cell surface antigens that frequently accompany the transformation of a cell. As a result, nearly all cancer cells are eliminated by the immune system before causing serious harm.

Agents that cause the transformation of a normal cell into a cancerous one are said to be *carcinogenic* or *oncogenic*. Among these agents are various chemicals, radiation, and infection by certain viruses. Carcinogens transform normal cells by directly or indirectly causing changes in the cell's hereditary material (i.e., DNA). These changes are called **mutations**, and the carcinogen is thus said to be *mutagenic*.

It is now clear that most (perhaps all) cancers stem from (1) changes in the chromosomal location of specific genes, (2) changes in the structure of specific genes, and/or (3) the activation of specific genes called **proto-oncogenes**. Proto-oncogenes are present in all normal cells, where they exist either in an inactive state or function in a normal way. Mutations change or activate proto-oncogenes, converting them to **oncogenes**, which then function abnormally.

Cancers of the Immune System

Burkitt's Lymphoma. Some cancers are clearly associated with changes in the locations of specific genes. Normal human somatic cells contain 46 chromosomes (i.e., 23 pairs), and each chromosome has a particular sequence of genes. Mutations that result in the movements of genes to other regions of a chromosome or to a different chromosome altogether may set the stage for the growth of tumors. One of the best illustrations of this is a human leukemia called **Burkitt's lymphoma**. In this disease, tumors appear in the B-lymphocyte-producing tissues.

In the most common form of Burkitt's lymphoma, a *reciprocal translocation* of a number of genes occurs between chromosome number 8 and chromosome number 14 (i.e., each chromosome can be distinguished from the others by microscopy and has been assigned a specific number). A segment at the end of chromosome 8 is translocated to chromosome 14 and is replaced by a segment from the end of chromosome 14. As a result, a proto-oncogene from chromosome 8 ends up near genes of chromosome 14 responsible for the production of antibody. This translocation alters the nature of the antibody that is produced and also serves to convert the proto-oncogene to an oncogene. The result is the excessive multiplication of the transformed cell creating the tumors that characterize this disease.

T-Cell Leukemia. It is now clear that a number of cancers result from infection by a class of viruses called **retroviruses.** The mechanisms by which the retroviruses cause cancer appear to be diverse and include (1) transferring oncogenes to the cells they infect (the retrovirus oncogenes are believed to have been picked up from previously-infected host cells), (2) activating oncogenes that are already present in the host cell, and (3) transferring genetic material to the host cells that causes the cells to undergo excessive multiplication.

Two retroviruses, called **HTLV-I** and **HTLV-II**, are the cause of **T-cell leukemias**. These leukemias are characterized by the excessive and abnormal production of T-cells. The HTLV viruses do not carry oncogenes (and therefore do not transfer oncogenes to the host) and also appear not to activate oncogenes already present in the host. One model proposed to explain the actions of the HTLV retroviruses proposes that the virus directs the infected host cell to produce a protein that stimulates not only production of new retroviruses but also production of new host cells. This leads to the abnormal and uncontrolled growth of the transformed cells.

A close relative of HTLV-I and HTLV-II, called **HTLV-III**, causes Acquired Immune Deficiency Syndrome (AIDS) (see earlier in the chapter). In contrast to the effects of HTLV-I and HTLV-II, HTLV-III causes T-cell death rather than excessive multiplication.

THE IMMUNE SYSTEM AND CANCER

As noted at the beginning of the chapter, in addition to protecting the body against infection, the immune system serves as a principal line of defense against cancer. Therefore, it is not surprising that suppression of the immune system often results in cellular transformation and the appearance of tumors. For example, it is not uncommon for individuals with AIDS to develop a cancer called **Kaposi's sarcoma.**

Tumor cells are not only functionally different from their normal predecessors but they may also be antigenically different. Tumor cells may lose the major histocompatibility antigens, may produce new antigens, or may produce antigens normally found only at much earlier (e.g., embryonic) stages of development. Tumors that stem from infection by oncogenic viruses may display antigens that are characteristic of the virus. If the antigens present in the surfaces of a tumor cell are sufficiently different from those of a normal cell, the cell will be regarded as "non-self" and will be attacked by the immune system.

THE RESPIRATORY SYSTEM

In this chapter, we will be concerned with the respiratory system. This organ-system is responsible for providing the tissues of the body with oxygen needed for metabolic processes that fuel the body's activities. The respiratory system also provides for the excretion of waste carbon dioxide produced during metabolism. As an auxiliary function, movements of the respiratory muscles are also essential for oral communication.

AN OVERVIEW OF RESPIRATION

The term **respiration** refers to the entire process by which the exchange of gases between the atmosphere that surrounds us and the cells that make up our tissues takes place. This involves (1) the transfer of oxygen to the blood from the atmosphere; (2) the transport of oxygen in the blood to the tissues; (3) the transfer of the oxygen from the blood to the tissue cells; (4) the transfer of waste carbon dioxide from the tissues to the blood; (5) the transport of carbon dioxide in the blood to the lungs; and (6) the excretion of the carbon dioxide into the surrounding atmosphere.

MAJOR ORGANS OF THE RESPIRATORY SYSTEM

The major organs of the respiratory system are depicted in Figure 10-1 and include the **nose** (and/or **mouth**), **pharynx, trachea, bronchi** and **bronchioles, lungs, diaphragm, ribs**, and **rib muscles** (i.e., the external and internal *intercostal* muscles). During inspiration, air passes through the nose (or through the mouth, if the nasal passageways are blocked), and down a system of channels within which the air is filtered, warmed, and moistened before reaching the lungs' air sacs. At the front of the **nasal cavity**, nasal hairs act to filter the larger air-borne particles. The nasal cavity and the respiratory passageways into which this cavity leads are lined with **ciliated, pseudostratified epithelium** (also known as **mucous membrane**; see Fig. 10-2). This tissue secretes a sticky substance called **mucus** that traps small debris. The hair-like **cilia** that project from the cells beat back and forth in synchrony, thereby sweeping the mucus and the entrapped particles toward the throat (where the mucus is swallowed). Some debris inevitably reaches the lungs,

153

but in the lungs **macrophages** ingest and destroy these particles.

The nasal and oral cavities are separated from one another by shelf-like structures called the **hard** and **soft palates**. At the rear of the soft palate, the oral and nasal cavities merge to form the **pharynx**. Air passes downward through the pharynx and into a tube called the **trachea** (also known as the "windpipe"). The trachea is kept open (even when swallowing) by cartilaginous rings. The upper end of the trachea forms the **larynx** (or voice box) which contains the **vocal cords**. The opening that leads into the larynx is protected by a muscular flap called the **epiglottis**. The epiglottis closes down over the entrance to the larynx whenever you swallow; as a result, food is prevented from entering the trachea and blocking the air passageway.

At its base, the trachea divides into two branches, the left and right **major bronchi**. Each major bronchus enters a lung, where it divides into successively smaller branches called **bronchioles**. The smallest of the bronchioles, called **respiratory bronchioles**, end as a cluster of small sacs called **alveoli** (Fig. 10-3). It is estimated that in the average person, the lungs contain about 300 million alveoli. The surface of each alveolus is composed of a thin layer of cells; on one side the surface faces air "drawn" into the lungs from the surrounding atmosphere. On the opposite surface, each air sac interfaces with an extensive network of capillaries formed from branches of the pulmonary arteries. Gases (e.g., oxygen and carbon dioxide) readily pass between the air of the air sacs and the blood of the alveolar capillaries by **diffusion** along concentration gradients.

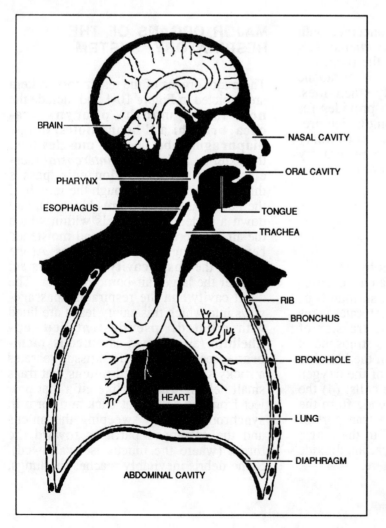

BRAIN

NASAL CAVITY

ORAL CAVITY

PHARYNX

ESOPHAGUS

TONGUE

TRACHEA

RIB

BRONCHUS

BRONCHIOLE

HEART

LUNG

DIAPHRAGM

ABDOMINAL CAVITY

Figure 10-1
The major organs of
the respiratory system.
See text for description.

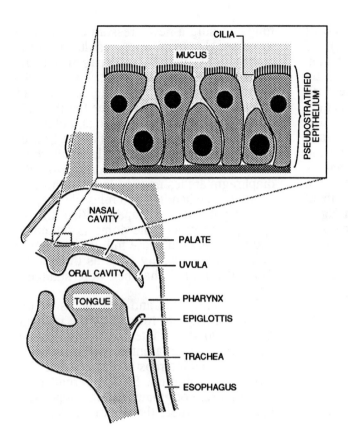

Figure 10-2
The nasal cavity is lined by **ciliated, pseudostratified epithelium** (shown in the rectangular insert). This tissue serves to filter, warm, and moisten the inspired air.

INSPIRATION AND EXPIRATION

Air and Atmospheric Pressure

The movements of air *into* and *out of* the lungs are respectively referred to as **inspiration** and **expiration**. These air movements are the consequences of changes in the relative pressures of the air in the atmosphere around us and the air that fills the lungs' air sacs. At sea level, the pressure of the air in the atmosphere is about 760 mm Hg. That is, the air in the atmosphere pushes down on the surface of the earth with a pressure equal to that exerted by a column of mercury that is 760 mm (or about 30 inches) high. The pressure varies somewhat from day to day (or hour to hour) depending upon atmospheric conditions and also diminishes rapidly with increasing altitude. So long as the nasal passageways are not blocked (or one's mouth is open), an equilibrium is maintained between atmospheric pressure and air pressure within the lungs. If atmospheric pressure exceeds the lung pressure, then air is pushed into the lungs (from the atmosphere) until a new equilibrium is attained. On the other hand, if the lung pressure exceeds atmospheric pressure, then air is pushed from the lungs into the atmosphere.

Air is a mixture of a number of different types of gases. The most abundant of the gases in air is *nitrogen* gas (i.e., N2), which accounts for close to 80% of all the gas molecules in air. Another 19-20% of the gas in air is *oxygen* gas (i.e., O2). Of the remaining gas, about 0.1% is *carbon dioxide* (i.e., CO2). As noted earlier, the pressure exerted on the surface of the earth by air (i.e., atmospheric pressure) is equal to that exerted by a column of mercury that is 760 mm high. The concentration of each of the gases in air is represented by the fraction of this total pressure that is represented by the gas. For example, the concentration of nitrogen gas in air is about 608 mm Hg (i.e., 80% of 760 mm Hg is 608 mm Hg); the concentration of oxygen is about 152 mm Hg (i.e., 20% of 760 = 152); and the concentration of carbon dioxide is about 0.8 mm Hg (0.1% of 760 = 0.8). Each of these values is a **partial pressure**.

Inspiration

Inspiration and expiration are the consequences of muscle contractions and relaxations that alter the air pressure within the lungs causing a new pressure equilibrium to be sought. Inspiration results from the contractions of the **diaphragm** and the **external intercostal muscles**. The diaphragm (see Fig. 10-1) is a thick sheet of striated muscle that separates the thoracic cavity (which houses the lungs and heart) from the abdominal cavity (which houses the stomach, small intestine, large intestine, and other organs). Prior to contraction, the diaphragm arches upward into the thoracic cavity (i.e., when it is in a relaxed state, the diaphragm is dome-shaped). When the diaphragm contracts, the dome is flattened, and this acts to increase the vertical height of the thoracic cavity (it is as though the floor of a sealed chamber were suddenly lowered).

The front, rear, and side walls of the thoracic cavity are formed by the **ribcage** and the two sets of intercostal muscles–the external and internal intercostal muscles. It is the external intercostal muscles that play a role in inspiration. Each external intercostal muscle originates on the upper surface of a rib where the rib articulates with

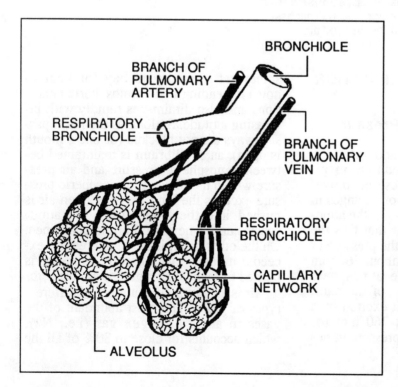

Figure 10-3
Each bronchus gives rise to a number of smaller **bronchioles**. The bronchioles, in turn, branch into **respiratory bronchioles**, which terminate in a cluster of **alveoli (air sacs)**. Each alveolus is covered by a capillary bed that receives blood from a branch of the pulmonary artery. Blood flowing through these capillaries is oxygenated prior to its return to the heart through the pulmonary veins.

the vertebral column (i.e., at the rear of the body). The muscle then runs around the ribcage and inserts onto the upper surface of the rib *below* where the rib joins the **sternum** (or breastbone). When the external intercostal muscles contract, they act to rotate the ribs upward (Fig. 10-4). This action pushes the sternum forward (further away from the vertebral column), thereby increasing the anterior-posterior dimensions of the thoracic cavity and increasing the cavity's volume (it is as though the front wall of a sealed chamber suddenly moved forward). Thus, the simultaneous contractions of diaphragm and external intercostal muscles act to enlarge the thoracic cavity. At rest, when one is breathing quietly, about 75% of the increase in volume of the thoracic cavity is due to the contraction of the diaphragm. The deeper (and more rapid) breathing that accompanies physical activity involves more extensive contractions of the external intercostal muscles.

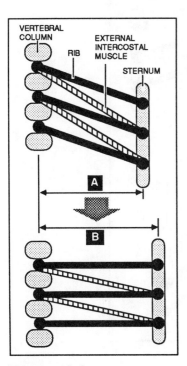

Figure 10-4
Contractions of the external intercostal muscles elevate the ribcage and push the sternum forward (from A to B). This increases the volume of the thoracic cavity.

Although the lungs are not attached to the walls of the thorax, the respective downward and forward excursions of the thoracic floor and front wall are accompanied by expansion of the lungs. Thus, the volume of the lungs increases in sympathy with the increased volume of the thorax.

The increase in lung volume that results from contractions of the diaphragm and external intercostal muscles drops the air pressure within the lungs to a value lower than atmospheric pressure. Consequently, air moves from the atmosphere into the lungs in order to bring the two pressures back into equilibrium. It is important to emphasize that air moves from the atmosphere into the lungs *because the lungs have expanded*. (The lungs are not expanded by the inward rush of air; rather, the expansion of the lungs causes air to be pushed into the respiratory passageways from the atmosphere. It is, therefore, not correct to infer that lung volume increases *because of the entry of air*. Recognizing the cause and effect relationships between changes in lung volume and movements of air between the atmosphere and the lungs are important to your understanding the mechanics of respiration.)

Expiration

Expiration, the movement of air from the lungs back into the surrounding atmosphere, can take either of two forms: **passive expiration** or **active expiration**. Passive expiration is the mechanism at work during resting or quiet breathing, whereas active expiration occurs when one is engaged in more strenuous activity (walking, running, etc.).

Passive expiration occurs when, at the end of inspiration, the diaphragm and external intercostal muscles are returned to their relaxed state. When the diaphragm relaxes, it resumes its normal shape and it once again arches upward from the abdominal cavity into the thoracic cavity. Relaxation of the external intercostal muscles is followed by the downward rotation of the ribcage. The downward movement of the ribs is assisted by gravity and by the natural elasticity of the thorax wall.

When the diaphragm arches upward into the thorax, the volume of the thorax is decreased (i.e., when the floor of the thorax rises, the chamber's vertical height is diminished). When the ribcage rotates downward, the sternum is pulled backward. This too acts to reduce the volume of the thorax (i.e., when the front wall of the thoracic cavity moves rearward, the chamber's depth is reduced). These movements exert pressure on the lungs causing the volume of the lungs to decrease. The decrease in lung volume raises the air pressure in the lungs to a value that exceeds atmospheric pressure. As a result of the pressure difference that now exists between air in the lungs and air in the atmosphere, air moves out of the lungs.

It is important to note that during expiration the decrease in lung volume *precedes* the movement of air out of the lungs. That is to say, air leaves the lungs *because the lung volume decreases* (i.e., the decreased lung volume raises the pressure, thereby causing air to move in the direction of the lower atmospheric pressure). (The lungs are not compressed by the outward rush of air; rather, the compression of the lungs causes air to be pushed out of the respiratory passageways into the atmosphere. It is, therefore, not correct to infer that lung volume decreases *because* of the loss of air.)

In the average person at rest, the cycle of inspiration and expiration that characterizes quiet breathing occurs about 16-18 times each minute. This rate is adequate to supply the body's needs for acquiring oxygen and eliminating carbon dioxide under resting conditions. During periods of increased muscular activity, both the rate and the depth of inspiration and expiration are increased. Increasing the depth and rate of inspiration is achieved by more forcible and more rapid contractions of the diaphragm and external intercostal muscles. However, the downward rotation of the ribcage that occurs when the external intercostal muscles relax is not rapid (or forceful) enough to provide the necessary rate and depth of

expiation. The *active expiration* that is required under these conditions is achieved as follows.

Attached to the upper surfaces of the ribs where the ribs articulate with the vertebral column are the origins of the **internal intercostal muscles**. Each of these muscles encircles the ribcage and inserts into the undersurface of the rib *above*, where the rib joins the **sternum**. When the internal intercostal muscles contract, they pull the ribs downward (Fig. 10-5). This action draws the sternum backward (closer to the vertebral column), thereby quickly decreasing the anterior-posterior dimensions of the thoracic cavity and decreasing the

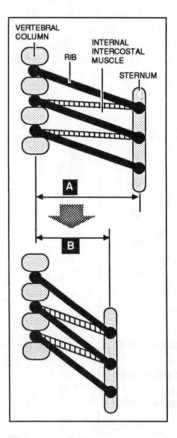

Figure 10-5

Contractions of the internal intercostal muscles pull the ribcage dowward drawing the sternum backward (from A to B). This quickly decreases the volume of the thoracic cavity.

cavity's volume. The actions of the internal intercostal muscles may be accompanied by contractions of the wall of the abdominal cavity (which lies below the thoracic cavity). Such contractions force the abdominal organs upward against the undersurface of the diaphragm causing the diaphragm to arch further into the thoracic cavity.

GAS EXCHANGES BETWEEN THE AIR SACS, THE BLOOD, AND THE TISSUES

Oxygenation of the Blood

As noted earlier, the concentration of oxygen in inspired air is about 152 mm Hg (i.e., the concentration of a gas is expressed as its partial pressure). An analysis of the oxygen concentration of the air in the lung's air sacs reveals that it is about 105 mm Hg. The lowered partial pressure of oxygen in alveolar air is due to the added water vapor content (radiated from the mucous membrane lining the respiratory passageways) and the increased carbon dioxide. By comparison, the concentration of oxygen gas dissolved in the blood plasma and red blood cells entering the pulmonary circulation is about 40 mm Hg. (An equilibrium exists between plasma and red blood cells such that each of the two phases contains the same concentration of dissolved oxygen.) The 65 mm Hg partial pressure difference between alveolar oxygen and pulmonary blood oxygen (i.e., 105 minus 40) causes oxygen gas to diffuse from the air sacs into the bloodstream (see Figs. 10-6 and 10-7).

Oxygen gas that diffuses into the blood from the alveoli dissolves in the blood plasma and also enters the red blood cells. If there were no hemoglobin in the red blood cells, then the blood leaving the lungs and returning to the left side of the heart would contain about the same amount of oxygen as alveolar air. However, the presence of hemoglobin dramatically alters the blood's oxygen-carrying capacity. Oxygen entering the red blood cells from

the plasma reacts with hemoglobin and becomes chemically bound to it. (Each hemoglobin molecule contains 4 atoms of iron and each of these iron atoms can bind two atoms of oxygen; see Chapter 8.) The oxygen that is bound to hemoglobin is no longer in the gaseous state and therefore *does not contribute to the oxygen partial pressure*. Consequently, the partial pressure gradient that causes oxygen to diffuse into the blood from the alveoli remains intact as hemoglobin accepts a full oxygen load. It is only when hemoglobin's oxygen binding capacity approaches the saturation point that the partial pressure of oxygen gas in the blood starts to approach the 105 mm Hg level present in the air sacs. The upshot of this is that hemoglobin raises the oxygen carrying capacity of the blood about 70-fold.

Were there no hemoglobin in red blood cells, then 100 c.c. of blood could accept

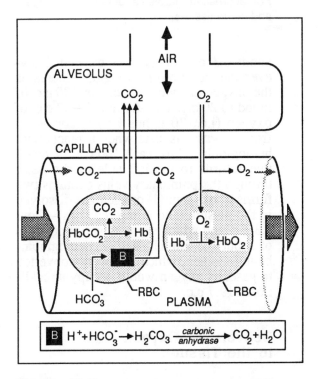

Figure 10-6
Exchanges of oxygen and carbon dioxide between the air sacs of the lungs and the pulmonary blood. See text for explanation.

159

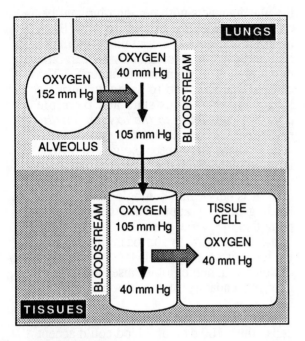

Figure 10-7
Partial pressure gradients for O2 in the lungs and in the tissues.

(about 40 mm Hg). Although some oxygen is lost to the walls of blood vessels carrying systemic blood away from the left side of the heart, the partial pressure of oxygen in the blood entering a tissue's capillary beds is not far below 100 mm Hg. The 60 mm Hg partial pressure difference between the blood and the tissue (i.e., 100 minus 40) results in the diffusion of oxygen out of the blood (Figs. 10-7 and 10-8). The falling partial pressure of oxygen in the blood acts as a trigger for the release of oxygen that has been bound to hemoglobin (Fig. 10-9). Upon its release from hemoglobin, the oxygen returns to its gaseous form, diffuses from the red cells into the plasma, and then passes along the sustained concentration gradient from the blood into the tissues. By the time blood leaves the capillary networks of a tissue and returns to the right side of the heart, the partial pressure of oxygen in the blood is about 40 mm Hg and about one-half of the oxygen that was bound to hemoglobin has been released.

about 0.3 c.c. of gaseous oxygen. However, the presence of hemoglobin increases the oxygen-carrying capacity of 100 c.c. of blood to 20 c.c. The additional 19.7 c.c. of oxygen (i.e., 20 minus 0.3) is converted from its gaseous state when it reacts with hemoglobin. Thus, each 100 c.c. of blood that returns to the left side of the heart through the pulmonary veins contains two forms of oxygen: (1) oxygen gas dissolved in the plasma (and in the red cells)–about 0.3 c.c. and which exerts a partial pressure of about 105 mm Hg, and (2) hemoglobin-bound oxygen, which if it were converted to gaseous oxygen would occupy an additional 19.7 c.c.

Delivery of Oxygen to the Tissues

Because they are continuously consuming oxygen during their metabolism, tissue cells (and the tissue fluid that bathes the cells) have a low oxygen partial pressure

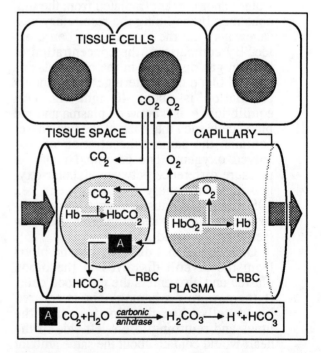

Figure 10-8
Exchanges of oxygen and carbon dioxide between the systemic blood and the tissues of the body. See text for explanation.

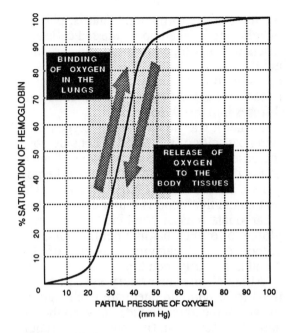

Figure 10-9
Oxygen/hemoglobin dissociation curve (see text for explanation)

Transport of CO2 in the Blood

Carbon dioxide is one of the major waste products of tissue metabolism. In a typical, actively metabolizing tissue, the partial pressure of CO_2 is about 50 mm Hg. In contrast, CO_2 dissolved in the systemic blood entering a capillary bed is about 40 mm Hg. The 10 mm Hg partial pressure difference (i.e., 50 minus 40) causes CO_2 to diffuse from the tissues into the bloodstream (Fig. 10-10). CO_2 entering the bloodstream from a tissue is carried back to the right side of the heart and then into the pulmonary circulation in three different chemical forms: (1) as a *dissolved gas*, (2) chemically *combined with hemoglobin*, and (3) as *bicarbonate ions* (Figs. 10-6 and 10-8).

About 7% of the CO_2 that enters the blood retains its gaseous form and dissolves in the plasma. Another 20% of the CO_2 diffuses into the red blood cells where it reacts with hemoglobin to form **carbamino hemoglobin**. Unlike oxygen, which combines with the iron atoms of

hemoglobin's heme groups, CO_2 combines with amino acids of this protein. (Do not confuse CO_2 with CO (i.e., *carbon monoxide*); CO is an extremely poisonous gas that displaces oxygen from hemoglobin and combines irreversibly with hemoglobin's iron atoms, thereby destroying hemoglobin's ability to transport oxygen.)

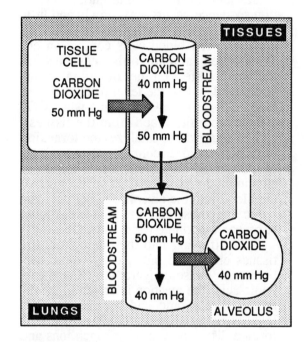

Figure 10-10
Partial pressure gradients for CO_2 in the lungs and in the tissues.

The remaining 73% of the carbon dioxide transported in the blood takes the form of **bicarbonate ions**, nearly all of which are carried in the plasma. The conversion of CO_2 to bicarbonate occurs in the following way. When carbon dioxide and water are mixed together, some of the CO_2 reacts with water to form carbonic acid; i.e.,

$$CO_2 + H_2O \xrightarrow{\text{Carbonic anhydrase}} H_2CO_3$$

Nearly all of the carbonic acid that is formed breaks down (i.e., "dissociates") immediately into hydrogen ions and bicarbonate ions, leaving only a small percentage of the carbonic acid molecules intact; i.e.,

$$H_2CO_3 \longrightarrow H^+ + HCO_3^-$$

In water (or in blood plasma), this reaction sequence occurs very slowly and little carbonic acid and bicarbonate are produced. However, red blood cells contain an enzyme called **carbonic anhydrase** that has a dramatic effect on the formation of carbonic acid. When the level of CO_2 in red blood cells is *rising* (as occurs when CO_2 diffuses into the blood from the tissues), carbonic anhydrase catalyzes the formation of carbonic acid. In contrast, when the level of CO_2 in red blood cells is *falling* (as occurs in the lungs when CO_2 diffuses out of the blood into the alveoli; see below), carbonic anhydrase catalyzes the breakdown of carbonic acid into water and CO_2.

CO_2 diffusing into the blood from the tissues enters red blood cells where carbonic anhydrase converts the CO_2 to carbonic acid (the reaction is driven in the direction of carbonic acid because the CO_2 level is rising). The carbonic acid that is formed dissociates into hydrogen ions and bicarbonate, and the bicarbonate is then secreted from the red cells into the plasma. Therefore, although the bicarbonate originates in the red blood cells, it is transported to the lungs in the blood plasma.

Elimination of CO_2 in the Lungs

The concentration of CO_2 gas in alveolar air (about 40 mm Hg partial pressure) is less than that in the blood entering the pulmonary capillaries (about 50 mm Hg partial pressure; Fig. 10-10). As a result, CO_2 diffuses from the blood into the air sacs. As the CO_2 level of the blood falls, hemoglobin-bound CO_2 is released, diffuses out of the red cells, and passes into the air sacs. Moreover, the falling level of CO_2 drives the carbonic anhydrase reaction in the direction of CO_2 formation. That is, plasma bicarbonate is taken up by the red cells and reacts with hydrogen ions to form carbonic acid, which the enzyme then quickly converts to CO_2 and water. The CO_2 produced in this way leaves the red cells and diffuses into the alveolar air.

Earlier it was noted that the oxygen-carrying capacity of the blood is considerably increased through the conversion of most of the dissolved gas to oxygenated hemoglobin. Because most of the carbon dioxide that enters the blood is either converted to bicarbonate ions or combines with hemoglobin to form carbamino hemoglobin, the carbon dioxide-carrying capacity of the blood is similarly increased.

Although the respiratory system functions to provide an adequate exchange of oxygen and carbon dioxide between the air and the tissues of the body, the major molecular component of air is nitrogen gas (N_2). Indeed, N_2 accounts for about 80% of all of the gas in air. Therefore, the air sacs of the lungs are rich in N_2. Nitrogen gas enters the bloodstream and establishes a partial pressure equilibrium with the nitrogen present in the alveolar air. However, nitrogen gas is not metabolized by the body and so none is consumed by the tissues. Blood contains no nitrogen carriers and nitrogen gas is not converted to any other molecular form. Consequently, the only N_2 in the blood is the small amount that dissolves in the plasma.

THE LUNG VOLUMES

The amount of air that enters and leaves the lungs during each respiratory cycle can be varied according to the body's need to acquire additional oxygen or eliminate more carbon dioxide. Changes in the volume of air entering and leaving the lungs are brought about by varying the extent and frequency of contractions of the diaphragm and external intercostal muscles during inspiration and by varying the actions of the internal intercostal muscles during expiration.

During quiet (i.e., resting) breathing,

air that enters (and leaves) the lungs in one respiratory cycle is called the **tidal volume**. In the average male this amounts to about 500 c.c. During maximal inspiration and expiration, the volume of air exchanged with the atmosphere can be increased about ten-fold (to about 5,000 c.c. or 5 liters). This maximum volume is called the **vital capacity** (Fig. 10-11). The additional air that can be inspired after a normal inspiration is completed is known as the **inspiratory reserve volume**. In a person whose vital capacity is 5,000 c.c., the inspiratory reserve volume amounts to about 3,500 c.c. The additional volume of air that can be expired after a normal expiration is completed is known as the **expiratory reserve volume**. In a person whose vital capacity is 5,000 c.c., the expiratory reserve volume amounts to about 1,000 c.c. As seen in Figure 10-11, the larger the tidal volume, the smaller are the inspiratory and expiratory reserve volumes.

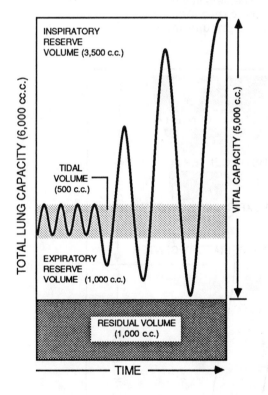

Figure 10-11
The various lung volumes in an average male.

Even after all of the expiratory reserve volume has been expelled from the lungs, the lungs are still not empty. About 1,000 c.c. of air remains in the lungs; this volume is known as the **residual volume**. The sum of the residual volume and the vital capacity is known as the **total lung capacity**. In a person whose vital capacity is 5,000 c.c., the total lung capacity amounts to 6,000 c.c.

REGULATION OF BREATHING

Although we rarely exercise conscious control over the respiratory movements, the intercostal muscles and diaphragm do not possess the spontaneous rhythmicity that characterizes heart muscle. Rather, the respiratory musculature must be stimulated by nerve impulses carried by certain spinal nerves. The actions of these nerves are regulated by respiratory centers in the brainstem.

Inspiratory and Expiratory Centers

In the medulla of the brainstem are two centers that play important parts in the regulation of breathing. These are known as the **inspiratory center** and the **expiratory center** (Fig. 10-12). Motor impulses initiated by the inspiratory center are carried to the external intercostal muscles by the **intercostal nerves** and to the diaphragm by the **phrenic nerves**. These impulses cause the external intercostal muscles and diaphragm to contract, thereby increasing the volume of the lungs and causing air to enter them. During resting breathing, impulses arising in the expiratory center act to inhibit the inspiratory center, in effect temporarily shutting this center off. In the absence of impulses reaching them, the external intercostal muscles and diaphragm relax and expiration ensues. During periods of greater activity, the expiratory center stimulates the internal intercostal muscles, thereby accelerating expiration.

The actions of the inspiratory and expi-

163

ratory centers may be augmented by a feedback mechanism (known as the Hering-Breuer reflex) in which sensory information reaches the brainstem from interoceptors in the lungs that sense the amount of stretching that takes place as the lungs expand.

Apneustic and Pneumotaxic Centers

Respiration is also influenced by the activities of two other respiratory centers in the brainstem; these are the **apneustic** and **pneumotaxic** centers, both of which reside in the **pons**. The apneustic center acts to stimulate the inspiratory center, promoting inspiration. However, the apneustic center is cyclically inhibited by the pneu-

motaxic center, so that the flow of stimulatory impulses being sent to the inspiratory center is temporarily interrupted permitting expiration to follow.

Influences of Blood Chemistry on Breathing

During vigorous activity, the body needs to consume more oxygen and produces more carbon dioxide. Oxygen deficiency and carbon dioxide excess act to increase both the depth and rate of respiration through two mechanisms: (1) direct effects of these two gases on the brain, and (2) indirect effects on the brain that are facilitated by peripheral chemical receptors (Fig. 10-12).

Figure 10-12
The depth and rate of breathing are regulated by nerve centers in the brainstem. These centers include the **inspiratory** and **expiratory centers** located in the medulla and the **apneustic** and **pneumotaxic** centers of the pons.

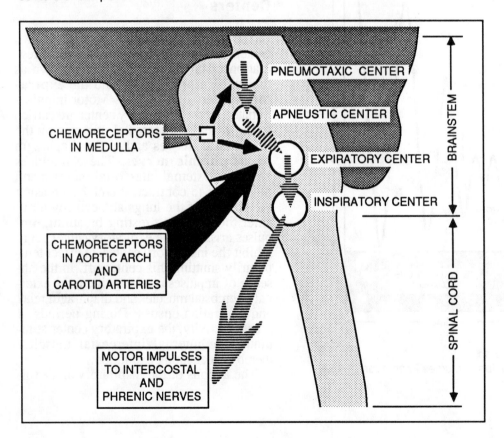

Peripheral Chemoreceptors. The aortic arch and carotid arteries contain interoceptors that respond to changes in the levels of oxygen and carbon dioxide dissolved in the blood (recall that in Chapter 7, it was noted that these structures also contain receptors that respond to changing blood pressure). A drop in the blood's oxygen level (or a rise in the carbon dioxide level) is followed by the activation of sensory nerve fibers that conduct impulses to the respiratory centers in the medulla; this is followed by an increase in motor impulses from the inspiratory center bringing about an increase in the rate and depth of breathing.

Medullary Chemoreceptors. The medulla contains chemoreceptors that are especially sensitive to the level of CO_2 dissolved in the blood. A rise in the CO_2 level (such as that which accompanies vigorous activity) is sensed by these receptors. As a result, impulses are relayed to the inspiratory center bringing about an increase in the rate and depth of breathing.

It has been shown the brain is more sensitive to rising levels of CO_2 in the blood than it is to falling levels of O_2. Carbon dioxide is, therefore, a greater stimulus for respiration than oxygen (this is why you are asked to breathe your expired air into the mouth of an unconscious person when applying mouth-to-mouth resuscitation).

SPEECH

The organs of the respiratory tract are not only responsible for gas exchanges between the atmosphere and the blood but also make speech possible. Speaking requires major modifications of the respiratory cycle (i.e., you cannot breathe in a normal, cyclical pattern and speak at the same time). The nerve impulses that provide for speech and that simultaneously modify the respiratory pattern arise in the cerebral cortex. These impulses go to the respiratory centers in the medulla and pons and to the musculature of the mouth, tongue, and larynx. Speech occurs during expiration as air is forced past the vocal cords causing them to vibrate. The frequency of vibration (hence the pitch of the sound) varies according to the position and tension in the vocal cords, properties that are regulated by the muscles of the larynx. Movements of the tongue and lips further modify the resulting sounds.

THE DIGESTIVE SYSTEM

The organs of the digestive tract break down the food that we eat and convert it into a form that can then be absorbed into the bloodstream and into the lymph. For the most part, the breakdown of the food is a chemical process in which digestive enzymes convert large molecules of proteins, carbohydrates, and fats into simpler, smaller molecules. The breakdown of the food is also assisted by a variety of mechanical processes, such as chewing and the vigorous muscular contractions of the walls of the digestive organs. Once digestion and absorbtion are completed, body water added to the food during its passage from one organ to the other is removed from the residual matter and returned to the bloodstream, with the result that a semi-solid waste (called **feces**) is eliminated from the body.

MAJOR ORGANS OF THE DIGESTIVE SYSTEM

The digestive system includes the digestive tract itself and the accessory organs and glands that provide the enzyme-rich digestive fluids (Figs. 11-1 and 11-2). The organs of the tract include (1) the **mouth** (or oral cavity), (2) the **pharynx**, (3) the **esophagus**, (4) the **stomach**, (5) the **small intestine**, (6) the **large intestine** (or **colon**), and (7) the **rectum**. Among the accessory digestive organs are (1) the **salivary glands**, (2) the **pancreas**, (3) the **liver**, and (4) the **gall bladder**. With the exceptions of the mouth, salivary glands, pharynx, and esophagus, the organs of the digestive system are confined to the body's abdominal cavity.

Although the food that we eat contains a great variety of different substances, the major food constituents are (1) **proteins**, (2) **carbohydrate** (much of it **polysaccharide**), and (3) **fat** (or **lipid**). The goal of the digestive system is to sequentially convert the protein into **amino acids**, the carbohydrate into **sugars**, and the fats into **sterols, glycerol, monoglycerides**, and **fatty acids**. The various stages through which these food materials pass during digestion are summarized in Figure 11-3.

THE MOUTH

The digestion of food begins almost immediately upon its entry into the mouth. The presence of food in the mouth serves as a chemical stimulus for receptors in the

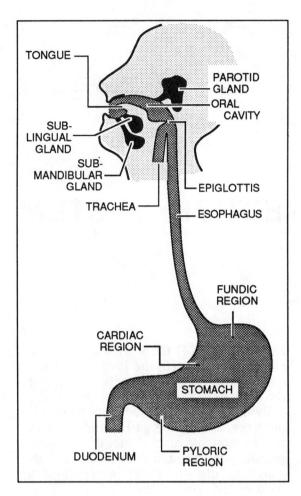

Figure 11-1
The upper region of the digestive tract.

tongue (the "taste buds" or gustatory receptors) which send sensory impulses to the central nervous system via associated sensory nerve fibers. This is reflexively followed by the passage of motor impulses from the brainstem to the three **salivary glands** (i.e., the *parotid*, *sublingual*, and *submandibular* glands) in the walls of the oral cavity bringing about a more rapid release of **saliva** into the food. Chewing, which serves to break the solid food into smaller pieces and mix it with saliva, also reflexively facilitates the release of saliva.

The major constituent of saliva is water, which serves to extract small molecules from the solid food and dissolve them. Since the gustatory receptors of the tongue

respond only to substances dissolved in water, dissolution of food materials is fundamental to the sense of taste. Also present in the saliva is **mucus**, a sticky material that serves to keep the shredded food together in one mass (called a **bolus**), so that it can be sliced and crushed more effectively during chewing. Saliva also contains a very important digestive enzyme called **salivary amylase** (also known as **ptyalin**).

Salivary amylase is an enzyme that can degrade polysaccharide (carbohydrate) by breaking specific chemical bonds that link neighboring sugars of the polysaccharide together. Although the degradative action of salivary amylase is very rapid, complete digestion of the carbohydrate is rarely achieved before the food is swallowed. As a result, the partially degraded carbohydrate that passes from the mouth into the stomach consists of branched sugar chains of reduced length, but not individual sugar molecules.

A Word About the pH Sensitivity of the Digestive Enzymes

Like virtually all other enzymes of the body, the enzymes produced by the organs of the digestive system are proteins and are especially sensitive to the pH of the fluid into which they are secreted (see Chapter 2 for a discussion of pH). Some enzymes function at neutral pH (i.e., around pH 7), some at alkaline pH (above pH 7), and others at an acidic pH (below pH 7). If the pH of the fluid in which digestion is proceeding is altered, enzymatic digestion may be abruptly halted.

In the case of salivary amylase, this enzyme functions at the alkaline pH that exists in the saliva. This pH persists as food is swallowed and passed into the stomach. However, the secretion of acid into the stomach (see below) dramatically lowers the pH, and this serves to halt the action of salivary amylase. Whereas carbohydrate digestion by salivary amylase is halted, the low pH established in the stomach provides the acidic environment in which the sto-

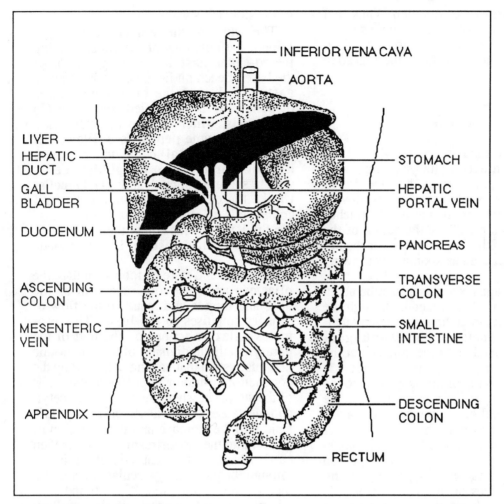

Figure 11-2
Major organs of the digestive tract located in the abdominal cavity. For clarity, most of the small intestine lying between the duodenum and the ileum has been removed.

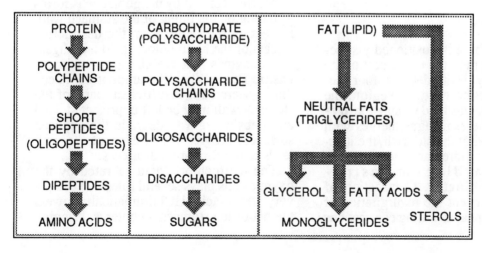
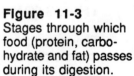

Figure 11-3
Stages through which food (protein, carbohydrate and fat) passes during its digestion.

169

mach's enzymes can function. Thus, pH changes that turn off some enzymes while turning on others are the norm as food passes from one segment of the digestive tract to another

THE ESOPHAGUS

The bolus of food is voluntarily pushed backwards into the pharynx and swallowed. This action drives the food downward into the esophagus. The swallowed food is prevented from entering the trachea (i.e., the "windpipe") by the action of the epiglottis, which closes down over the entry into the trachea as soon as swallowing begins.

The esophagus is a muscular tube about ten inches long that descends through the thoracic cavity, then through a small opening in the diaphragm (the **esophageal hiatus**), and leads into the stomach at the **gastro-esophageal junction**. Food is swept from the top of the esophagus to the stomach by wavelike contractions of the esophageal wall called **peristaltic waves**. These waves travel at about one inch per second, so that swallowed food remains in the esophagus for no longer than about ten seconds. Since the esophagus secretes no digestive enzymes and does not alter the pH of the swallowed material, the digestion of carbohydrate that began in the mouth is continued while the food travels through the esophagus.

THE STOMACH

The stomach, which is positioned just below the diaphragm in the upper part of the abdominal cavity, is divided into four successive regions: the **cardiac region**, the **fundic region**, the **body**, and the **pyloric region**. The esophagus merges with the cardiac region of the stomach, the junction protected by a circular band of muscle called the **cardiac** (or **gastro-esophageal**) **sphincter**. Contraction of this sphincter prevent the regurgitation of food into the esophagus during contractions of the stomach's walls.

The walls of the stomach are deeply folded, the openings of the folds forming the so-called **gastric pits** (Fig. 11-4). At the base of each pit lie the cells that secrete **gastric fluid** into the food. Four classes of cells may be identified; these are (1) **chief cells**, which secrete the stomach's digestive enzyme **pepsin** (see below), (2) **oxyntic** (or **parietal**) cells, which secrete **hydrochloric acid**, (3) **goblet cells**, which secrete mucus, and (4) **G cells**. Unlike the chief cells, oxyntic cells, and goblet cells whose secretions form the gastric fluid, the G cells empty their secretions (a hormone called **gastrin**) into the bloodstream.

The odor and taste of food stimulate the olfactory and gustatory receptors of the nose and tongue and result in the flow of sensory impulses to the brain. This promotes reflexive actions on the part of the parasympathetic division of the autonomic nervous system that are characterized by the conduction of motor impulses to the walls of the stomach, where the release of acetylcholine causes an initial secretion of gastric fluid into the stomach and the secretion of gastrin into the bloodstream. Gastrin is then carried back to the right side of the heart, through the pulmonary circulation, and into the systemic circulation. When blood containing gastrin flows through the capillary networks of the stomach's walls it promotes the more quantitative release of gastric fluid by the chief cells, oxyntic cells, and goblet cells.

Mucus secreted by the goblet cells forms a coating on the lining of the stomach wall, protecting the lining from the destructive effects of the hydrochloric acid and digestive enzymes that are also present in the gastric fluid. When the amount of mucus that is secreted is insufficient, parts of the stomach wall may be left unprotected. HCl and pepsin begin to degrade the exposed surfaces; progressive erosion of the wall leads to the formation of **ulcers**.

The hydrochloric acid secreted by the oxyntic cells plays several roles: (1) being a very strong acid, HCl dramatically lowers the pH of the stomach's contents, bringing

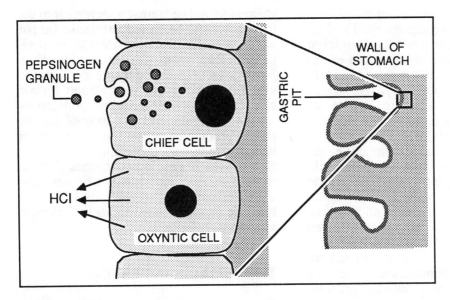

Figure 11-4
The walls of the stomach are deeply folded to form the **gastric pits** (seen on the *right*). The pits' **chief cells** secrete the protein-digesting enzyme **pepsin**, while the **oxyntic cells** secrete **hydrochloric acid**.

carbohydrate digestion to a halt and providing the acidic environment in which the stomach's enzymes can function; (2) since uncooked or inadequately cooked food frequently contains large numbers of microorganisms, the strong acid of the gastric secretion acts to kill most (but not all) of the microorganisms that are present; (3) HCl denatures protein present in the food, converting the protein to a form that is more easily digested by protein-digesting enzymes; and (4) HCl play a role in the activation of the stomach's digestive enzymes.

In adults, the stomach secretes a single digestive enzyme called **pepsin**. In acidic surroundings, pepsin is a powerful protein-digesting enzyme. However, when it is initially released from the stomach's chief cells, pepsin exists in an inactive form called **pepsinogen** (the inactive form of an enzyme is called a **proenzyme** or **zymogen**). Pepsinogen is activated (i.e., converted to pepsin) by the removal of small peptide fragments from the proenzyme. This is achieved by the action of HCl. Once a small amount of the active enzyme is formed, the enzyme itself serves to convert additional pepsinogen to pepsin. In a cascade-like series of reactions, the amount of activated enzyme present in the stomach increases rapidly and geometrically (Fig. 11-5).

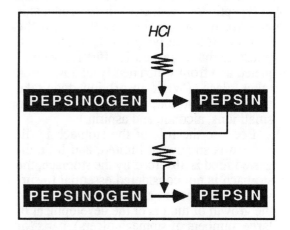

Figure 11-5
Activation of pepsin in the stomach. HCl promotes the conversion of some pepsinogen to pepsin, after which pepsin serves to activate additional pepsinogen molecules.

Peristaltic waves spreading across the walls of the stomach churn the food and act to mechanically break down food particles; the churning action also serves to effectively mix the food with the gastric fluid. In this manner, the food is progressively liquified forming what is known as **chyme**. The pyloric region of the stomach leads through the **pyloric valve** into the first segment of the small intestine (i.e., the duodenum). The stomach's peristaltic waves push the chyme against the pyloric valve, and with each wave a small amount of chyme is driven into the small intestine. The remainder of the chyme is refluxed back into the body of the stomach where it is subjected to additional churning and additional chemical degradation. Eventually all of the chyme is swept into the small intestine.

At this point, it might be well to ask how much digestion has been completed by the time the food reaches the small intestine? In fact, only a small amount of digestion is completed at this stage and is limited to the actions of salivary amylase on carbohydrate and the actions of pepsin on protein. There has been no fat digestion whatsoever. Salivary amylase reduces the carbohydrate in food to individual chains of sugars (polysaccharides), and pepsin reduces protein to short chains of amino acids (small polypeptides). Neither of these products is in a form that can be absorbed into the blood or into the lymph. It is in the small intestine that digestion is effectively completed and from which nearly all absorbtion takes place. (What absorbtion does take place in the stomach is limited to certain small ions, alcohol, and aspirin.)

Because the role of the stomach in digestion is somewhat limited, and little digested food is absorbed by the stomach, the stomach is not considered essential to survival. When necessary (e.g., as a result of the growth of tumors or the development of large numbers of stomach ulcers), parts (or all) of the stomach may be surgically removed (an operation called a *gastrectomy*) without disastrous consequences. There is, however, one vital function of the stomach that must be taken into account following a gastrectomy. The stomach secretes into the food a substance called **intrinsic factor** which is essential for the absorbtion of **vitamin B12** from the small intestine. Intrinsic factor secreted by the stomach and vitamin B12 present in the food form a complex that is later absorbed from the intestine. Vitamin B12 is essential for the proper production of red blood cells in the bone marrow; inadequate uptake of vitamin B12 causes **pernicious anemia**. Therefore, to survive, individuals who have had a gastrectomy must receive oral doses of intrinsic factor or intravenous injections of vitamin B12.

THE SMALL INTESTINE

There are two organs in the abdominal cavity referred to as *intestines*; these are the **small intestine** and the **large intestine**. The words "small" and "large" refer to the diameter of the intestine, not the intestine's length. Thus, the small intestine is a narrow tube, whereas the large intestine is a wide tube. Insofar as the lengths of these organs are concerned, the small intestine is very much longer than the large intestine. (The large intestine is commonly referred to as the **colon**.) It is the small intestine into which food passes from the stomach.

The walls of the small intestine are rich in smooth muscle. In a living person, this musculature is always contracted to some extent, so that the length of the organ rarely exceeds 12 feet. In a cadaver, the musculature is no longer contracted so that the small intestine extends to its maximum length. The relaxed length of the small intestine is more than 20 feet.

The first foot of length of the small intestine is called the **duodenum**. The remaining length is approximately equally divided into the **jejunum** and the **ileum**. The ileum leads into the large intestine (colon). It is in the small intestine that the digestive process is effectively completed, and it is from the small intestine that nearly all of the digested food is removed by absorbtion into the bloodstream and into the lymphatic circulation.

Secretions of the Liver, Gall Bladder, and Pancreas

Chyme pushed into the duodenum from the stomach causes the walls of the duodenum to release two substances (i.e., *hormones*) into the bloodstream; these are **secretin** and **cholecystokinin** (abbreviated CCK). These substances circulate through the blood, and upon reaching the pancreas, liver, and gall bladder promote the release of **bile** and **pancreatic juice**, which are conveyed by a series of ducts into the duodenum (Figs. 11-2 and 11-6). Consequently, the partially-digested food driven into the duodenum by contractions of the stomach's walls is almost immediately mixed with bile and pancreatic juice.

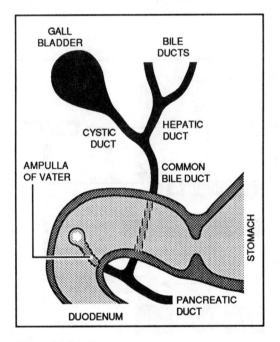

Figure 11-6
Relationships among the ducts that lead from the liver, gall bladder, and pancreas into the duodenum.

The Liver. The liver continuously produces and secretes bile, although secretion is augmented when secretin and CCK reach the liver from the duodenum. Bile, which leaves the liver through the **bile ducts** and

hepatic duct, is a watery solution containing a variety of inorganic and organic substances. Especially important are the **bile salts.** Bile salts are not salts in the conventional sense (such as sodium chloride, potassium chloride, etc.); rather these are large organic molecules. Upon reaching the small intestine and mixing with the chyme, the bile salts act to **emulsify** fat that is present in the food. Typically, the fat present in chyme takes the form of large droplets called **lipid droplets.** The bile salts emulsify the fat in an action much like that performed by a detergent. The bile salts aggregate at each lipid droplet's surface, breaking the droplet into successively smaller fragments. The smallest fragments can then be acted on by lipid-digesting enzymes that are also released into the small intestine (see below).

Also present in bile is **bile pigment** (or **bilirubin**). This colored substance is produced by the reticulo-endothelial tissues of the spleen, bone marrow, and liver during the breakdown of hemoglobin (bilirubin consists of a modified heme group that also has lost its central iron atom). Bilirubin produced in the bone marrow and spleen is shuttled to the liver by the plasma protein albumin. Bilirubin collected in the liver and produced by the liver itself is secreted in the bile and makes its way into the small intestine. Bacteria living in the small intestine convert the bilirubin to another pigment called **urobilinogen.** Thus, it is urobilinogen that colors the feces. Some of the urobilinogen produced in the small intestine is absorbed into the blood along with digested food. Reabsorbed urobilinogen eventually is removed from the blood by the kidneys and is excreted from the body in the urine. Consequently, the color of urine (as well as feces) is due to its content of the breakdown products of hemoglobin.

The Gall Bladder. As noted above, the liver produces and secretes bile constantly, not just during digestion. Between meals, bile leaving the liver through the bile ducts and hepatic duct does not pass directly into the duodenum. Instead, it is directed through the **cystic duct** into the **gall**

bladder. In the gall bladder, much of the bile's water is removed and returned to the bloodstream, and mucus is added. When CCK that is secreted into the blood by the duodenum reaches the gall bladder, it causes the muscular walls of the gall bladder to contract. Thus, the thickened and concentrated bile is pumped out of the gall bladder, through the cystic duct and **common bile duct**, toward the duodenum (Fig. 11-6).

The Pancreas. The pancreas produces and secretes into the duodenum a variety of substances that are essential to digestive activity. These substances include **bicarbonate**, which raises the pH of the chyme and inactivates pepsin, and several powerful digestive enzymes. The major enzymes produced by the pancreas are **pancreatic amylase**, **trypsin**, **chymotrypsin**, **carboxypeptidase**, and **pancreatic lipase**; these enzymes are most effective at the alkaline pH created by the bicarbonate that is mixed with the chyme. The secretion of pancreatic juice is stimulated by CCK and secretin. Pancreatic juice leaves the pancreas through the **pancreatic duct** and enters the duodenum, along with the bile, through the **ampulla of Vater** (Fig. 11-6).

Pancreatic amylase digests carbohydrate by degrading polysaccharide chains into successively smaller pieces. Consequently, although the digestion of carbohydrate is temporarily halted in the stomach, it is resumed in the small intestine. Trypsin, chymotrypsin, and carboxypeptidase are protein-digesting enzymes that continue the enzymatic breakdown of protein begun by pepsin. These enzymes continue to cleave polypeptide chains into smaller and smaller fragments. Carboxypeptidase attacks the *ends* of polypeptide chains, whereas trypsin and chymotrypsin attack the bonds that link amino acids together *within* a polypeptide chain. Pancreatic lipase is a fat-digesting enzyme that attacks the tiny lipid droplets produced by the emulsifying action of the bile salts.

Like pepsin, trypsin, chymotrypsin, and carboxypeptidase enter the duodenum in an inactive (i.e., zymogen) form and must be activated in order to resume the enzymatic digestion of protein in the chyme. The zymogen forms of these enzymes are called **trypsinogen**, **chymotrypsinogen**, and **pro-carboxypeptidase**. The activation of these zymogens takes the following course (Fig. 11-7). Some of the trypsinogen entering the duodenum is converted to trypsin by **enterokinase**, an enzyme released into the chyme from the mucosal lining of the duodenum. The trypsin formed in this manner then activates additional trypsinogen molecules (just as pepsin activates pepsinogen). Consequently, all of the trypsinogen is quickly converted to trypsin. Chymotrypsinogen and pro-carboxypeptidase are also converted to their active forms by trypsin (Fig. 11-7).

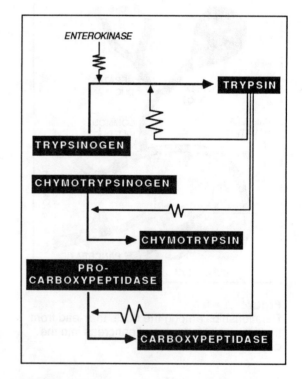

Figure 11-7
Conversion of the zymogens trypsinogen, chymotrypsinogen, and pro-carboxypeptidase to their active enzyme forms, trypsin, chymotrypsin, and carboxypeptidase.

Chyme entering the duodenum from the stomach is a very thick (pasty) substance. However, the secretions of the liver and pancreas include very large volumes of water, so that the chyme is quickly diluted as it passes on toward the jejunum and ileum.

Digestion in the Small Intestine

The walls of the small intestine contain large amounts of smooth muscle arranged in two patterns. Some of the smooth muscle runs around the circumference of the intestine and serves to constrict and dilate different regions of the intestine's length. The constrictions occur simultaneously at intervals along the intestine, thereby dividing the intestine's length into a number of consecutive chambers and acting to mix the chyme with the digestive enzymes. The narrow openings that interconnect successive segments are called **valves of Kerckring** (Fig. 11-8). Other smooth muscle runs longitudinally through the intestinal wall (i.e., this smooth muscle is oriented parallel to the long axis of the intestine). Contractions of this muscle are responsible for the peristaltic waves that slowly drive the chyme forward.

The inside surface of the small intestine is not smooth. Instead, millions of tiny finger-like projections extend from the surface into the intestinal lumen. These projections are called **villi** (*singular:* **villus**; Fig. 11-9), and they greatly increase the surface area of the intestinal wall. The increased surface area maximizes the organ's absorptive properties.

The villi are lined at their surface by epithelium consisting of two types of cells: **goblet** cells and **digestive/absorptive** cells. The goblet cells secrete mucus which helps to protect the intestinal lining from the actions of the digestive enzymes. Mucus is also produced and secreted by the **crypts of Lieberkuhn** (Fig. 11-9), small glandular structures in the wall of the intestine between neighboring villi. The digestive/absorptive cells produce the **intestinal enzymes** which act to complete the digestion of the food, converting it to a form that can be absorbed into the blood and lymph. These digestive enzymes are not secreted into the lumen of the intestine. Rather they are anchored in the surfaces (i.e., plasma membranes) of the thousands of **microvilli** that project from each cell. (Don't confuse the terms "villi" and "microvilli." Villi are the numerous projections from the wall of the intestine, each villus covered by hundreds of epithelial cells. Microvilli are microscopic projections from the surfaces of individual epithelial cells covering each villus.) Collectively, the microvilli form what is called a **brush border** (Fig. 11-10).

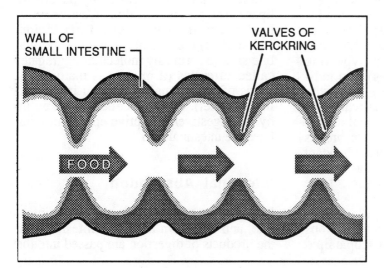

WALL OF SMALL INTESTINE

VALVES OF KERCKRING

FOOD

Figure 11-8
The length of the small intestine is divided into consecutive segments by constrictions of the intestinal wall. The narrow openings that interconnect the segments are called **valves of Kerckring**.

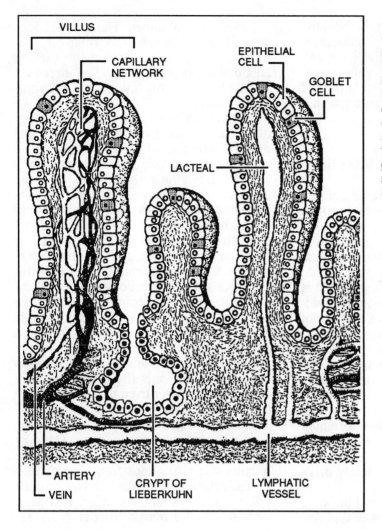

Figure 11-9
Organization of the villi that cover the surface of the small intestine. Each villus contains both a capillary network and a lacteal. Digestive/absorptive epithelial cells line the surface of each villus; scattered among the digestive/absorptive cells are mucus-secreting **goblet cells**. Mucus is also secreted by the **crypts of Lieberkuhn**.

The intestinal enzymes bring the digestive process to completion. Among the intestinal enzymes are **aminopeptidase, maltase, sucrase**, and **lactase**. Aminopeptidase breaks the short polypeptide fragments (i.e., **oligopeptides**) into individual amino acids. Maltase cleaves the **disaccharides** produced by salivary and pancreatic amylase into individual sugars (i.e., **monosaccharides**). Lactase breaks the disaccharide *lactose* (which is abundant in milk) into individual sugars, and sucrase breaks the disaccharide *sucrose* (i.e., "table sugar") into individual sugars.

The emulsification of lipid that is achieved in the chyme through the actions of the bile salts produces microscopic lipid droplets that can be digested by pancreatic lipase. The products of this digestive activity (e.g., **monoglycerides, fatty acids,** and **sterols**) then combine with additional bile salts to form tiny molecular aggregates called **micelles** (a micelle is many thousands of times smaller than an individual lipid droplet). These micelles are taken up by the digestive/absorptive epithelium that lines the intestinal villi.

Intestinal Absorbtion

Digestion is completed in the small intestine, and it is from the small intestine that the products of digestion are passed into the blood and the lymph that circulate through each intestinal villus. As seen in Figure 11-9, each villus contains a capillary network formed by branches of the **mesenteric arteries** that carry blood into the walls of the intestine. Each villus also contains a blind-ended branch of the lymphatic system that courses through the intestinal wall. These bulb-like structures which rise up through the center of each villus are called **lacteals**. Within each villus, the capillary network is wound around the central lacteal (Fig. 11-9).

Absorbtion of Amino Acids and Sugars

Amino acids produced by the digestion of protein are actively taken up by the digestive/absorptive epithelium lining the villi. From here the amino acids pass into the

blood that is circulating through each villus' capillary network. Although nearly all protein in food is converted to amino acids before it is absorbed, there are a few proteins that can be absorbed intact. For example, the intestinal epithelium of a newborn baby can absorb undigested antibody molecules. Since antibodies are present in the milk produced by the mother's mammary glands, a degree of immunity is conferred to a young baby that is breast feeding. The uptake of whole proteins by intestinal epithelium involves an entirely different cellular process than that involved in the uptake of free amino acids.

Sugars produced by the digestion of carbohydrate are also actively taken up by the digestive/absorptive epithelium. Like the amino acids, these sugars (principally glucose) are then passed into the blood circulating through the villus' capillary network.

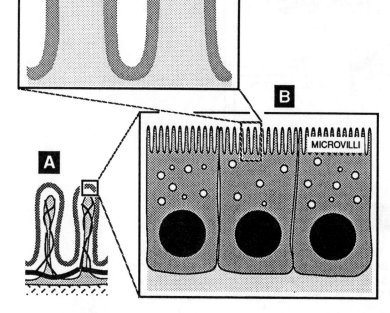

Figure 11-10
The intestine's villi are covered by digestive/absorptive epithelium (A and B). The exposed surfaces of the epithelial cells form numerous finger-like projections called **microvilli** (B and C). Anchored in the microvilli are the intestinal digestive enzymes (C).

INTESTINAL DIGESTIVE ENZYMES ANCHORED IN SURFACE OF THE MICROVILLI

Absorbtion of Lipid

Lipid micelles produced in the small intestine are taken up intact by the intestinal epithelium. Within the epithelial cells, the micelles are disassembled and their lipid is combined into larger and larger droplets. The droplets are then covered by protein synthesized by the epithelial cells, thereby forming small particles called **chylomicrons**. The chylomicrons are then transferred from the epithelium to the lacteals at the center of the villi. *(Note that lipid is transferred into the lymph whereas amino acids and sugars are transferred to the blood.)*

THE HEPATIC PORTAL SYSTEM

The fact that sugars and amino acids are passed into the blood, whereas lipids are passed into the lymph, has some very important implications. As illustrated in Figure 11-11, blood leaving the wall of the small intestine via the **mesenteric veins** does not return directly to the right side of the heart. Instead, this blood is shunted into the **hepatic portal vein** which carries the blood into the liver. This arrangement in which systemic blood flows through *two successive organs* (and their capillary beds) is an exception to the general rule that systemic blood flows through a single organ during each cycle through the systemic circulation. The implication of this arrangement is that the liver has an opportunity to screen the composition of blood leaving the intestine before that blood enters the general circulation.

As a result of digestion and absorbtion, the chemical composition of blood leaving the wall of the intestine is dramatically altered; for example, the blood's content of sugars and amino acids is markedly elevated. The sugar- and amino acid-rich blood is directed via the mesenteric and hepatic portal veins into the liver, where sugars and amino acids are removed from

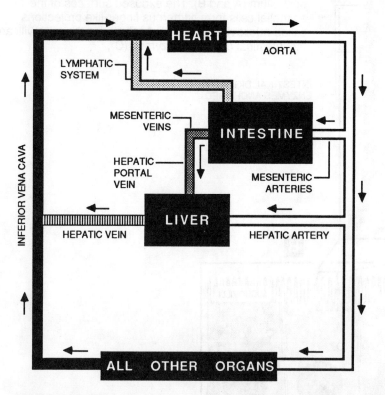

Figure 11-11
The **hepatic portal system** is an exception to the general rule that systemic blood passes through only one organ before returning to the right side of the heart. Blood leaving the walls of the small intestine is directed through the liver before returning to the heart.

the blood and temporarily stored. Between meals, as the various tissues of the body require additional sugars and amino acids for their metabolism, stored sugar and amino acids are released into the blood by the liver in order to meet tissue demands. In this way, the actions of the liver ensure that the chemical composition of the blood in the general circulation does not fluctuate dramatically.

Although it may be rich in sugars and amino acids, the blood entering the liver through the hepatic portal vein is poor in oxygen (having given up oxygen to the walls of the small intestine). The oxygen needs (and other needs) of the liver are met by the **hepatic artery** (a direct branch of the aorta, see Fig. 11-11) which carries fresh systemic blood to the liver. Portal and systemic blood circulating through the liver are carried out of this organ by the **hepatic vein,** which directs the blood into the inferior vena cava and then the right atrium.

In sharp contrast to the fate of absorbed sugars and amino acids (which are passed into the blood), the lipid-rich chylomicrons are passed into the lacteals and are carried away from the intestine with the lymph. This lymph is carried upward through the abdominal cavity into the thoracic cavity by lymphatic vessels. Within the thoracic cavity, the **thoracic duct** empties the chylomicron-rich lymph into the blood of the **left brachiocephalic vein** (see Chapter 7). From there the chylomicrons enter the right atrium and become part of the general circulation. Consequently, beginning near the heart, all tissues of the body have immediate access to the lipids of the chylomicrons. This creates a potentially harmful scenario in which lipids are removed from the blood and accumulate in excess in the walls of blood vessels near the heart and in the muscle tissues of the heart itself.

THE LARGE INTESTINE (COLON)

Chyme passes into the large intestine from the ileum of the small intestine and begins to make its way through the large intestine's three major sections: the *ascending colon* (on the right side of the abdominal cavity), the *transverse colon* (oriented horizontally, just below the stomach and pancreas), and the *descending colon* (on the left side of the abdominal cavity; see Figure 11-2). The surface of the colon is lined by epithelial cells, some of which secrete mucus that serves as a lubricant for the expulsion of the feces. Also present are crypts of Lieberkuhn; there are, however, no villi. During its passage through the colon, the chyme is altered in several ways and ultimately forms the semi-solid feces that are eliminated from the body through the **rectum** and **anus**.

Absorbtion of Water From the Colon

During digestion the small intestine contains large quantities of water. The water is derived from the ingested food and by the secretions of the organs of the digestive tract. It is estimated that in the average person, about 10 liters of water enter the small intestine daily–about 2 liters are ingested with the food and the remaining 8 liters are secreted by the digestive organs (especially the pancreas).

About 80% of the water in the small intestine is absorbed into the blood along with amino acids and sugars (and other substances) produced by the digestive process. The remaining 20% (about 2 liters per day) passes into the large intestine. Of this amount, about 90% (i.e., 1800 c.c. is absorbed from the colon into the blood), while the remainder (i.e., about 200 c.c.) is eliminated from the body with the feces.

Colonic Bacteria

A number of bacteria live and grow in the large intestine. These colonic bacteria live off the undigested and unabsorbed food that enters the colon from the small intestine. During their metabolism the colonic bacteria produce and excrete a variety of substances including several that are

179

essential to human metabolism. Included among the useful excretory products of the colonic bacteria are a number of vitamins (e.g., B12, K, thiamin, riboflavin). These vitamins are absorbed by the epithelial lining of the colon and transferred to the bloodstream.

The colonic bacteria also produce gaseous waste products including methane gas and hydrogen gas. Some of the gas is absorbed through the walls of the colon, but much of it exits the body through the anal opening along with the feces.

CHAPTER 12

THE EXCRETORY SYSTEM

Excretion is defined as the process by which the body rids itself of chemical substances (usually wastes) that were part of a tissue, the bloodstream, or the lymph. Accordingly, several different organs may be regarded as excretory organs. For example, the lungs could be considered excretory organs because they act to rid the body of carbon dioxide that is formed in the tissues during metabolism. The skin might be regarded as an excretory organ because the water and other chemical constituents secreted onto the body surface by the skin's sweat and oil glands were formerly in the tissues or in the bloodstream. Similarly, the liver might be regarded as an excretory organ because of its role in ridding the body of the breakdown products of hemoglobin (i.e., the bile pigments). Usually, however, the lungs are considered part of the respiratory system, the skin is considered part of the integumentary system, and the liver part of the digestive system. It is the two **kidneys** that represent the major excretory organs of the body. These vital organs serve to remove most of the chemical wastes emptied into the blood during metabolism, and working in concert with the urinary bladder, convey these wastes from the body.

The two kidneys are bean shaped organs about four inches long that lie on either side of the vertebral column in the lower part of the abdominal cavity (Fig. 12-1). The kidneys receive systemic blood from branches of the aorta called **renal arteries,** and blood leaving the kidneys enters the inferior vena cava through the **renal veins.** (Whenever the word "renal" is used in connection with an anatomical structure, it implies that the structure is related to the kidneys; for example, the adrenal glands ["ad" = "next to" and "renal" = "kidney" are the two endocrine glands seated on the upper surface of the kidneys; see Fig. 12-1].) About 1,000 c.c. (i.e., one liter) of blood enter the two kidneys each minute. Wastes are removed from this blood and form the **urine**. Approximately 2 c.c. of urine are formed each minute.

Extending from the kidneys downward to the **urinary bladder** are the two **ureters.** These tubes convey the urine from the kidneys into the urinary bladder, which acts as a temporary reservoir. Urine does not descend through the ureters by gravity. Rather, peristaltic waves (about four waves each minute) beginning at the upper, flared ends of the ureters progressively sweep the urine downward.

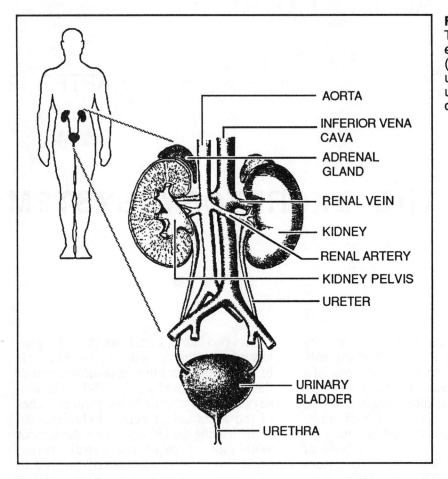

Figure 12-1
The major organs of the excretory system (kidneys, ureters, urinary bladder, and urethra). See text for discussion.

AORTA

INFERIOR VENA CAVA

ADRENAL GLAND

RENAL VEIN

KIDNEY

RENAL ARTERY

KIDNEY PELVIS

URETER

URINARY BLADDER

URETHRA

Periodically, the muscular walls of the urinary bladder contract and expel the urine from the body through a duct called the **urethra**. In males, the urethra is slightly longer than in females.

NEPHRONS

The 2 c.c. of urine that are formed from each liter of blood circulating through the kidneys is a product of the kidneys' functional units which are called **nephrons** or **kidney tubules**. Nephrons are microscopic structures, each kidney containing about one million of these units. An understanding of how the kidneys function is based on an appreciation of how the nephrons work. Since all nephrons function in an identical manner, we will focus on the actions of a single unit.

Organization of a Nephron

The structure of a nephron is depicted in Figure 12-2. Each tubule begins as a blind-ended, cup-shaped structure known as a **Bowman's capsule**. The capsule houses a capillary bed known as a **glomerulus**. Blood enters the glomerular capillary beds through small arterioles called **afferent arterioles** and leaves the capillary beds through **efferent arterioles** (i.e., "afferent" = "toward" and "efferent" = "away from"). (Note that the glomerular capillaries are drained by an arteriole and not a venule; the significance of this will become apparent later.) The Bowman's capsule leads into a twisted segment of the nephron called the **proximal convoluted tubule** ("proximal" means "nearby," so that the proximal convoluted tubule is nearby the Bowman's capsule). The proximal convoluted tubule leads into the **loop of Henle**, a U-shaped segment of the kidney tubule. The loop of Henle leads into the **distal convoluted tubule** (i.e., "distal" infers "distant," the distal convoluted tubule being the twisted segment of the nephron that is further from the Bowman's capsule). The distal convoluted tubule merges with a **collecting duct**; a single collecting duct receives the contents of a number of neighboring nephrons.

The efferent arteriole emerging from the Bowman's capsule gives rise to a second capillary network that weaves its way around the proximal convoluted tubule, loop of Henle, and distal convoluted tubule. This array of capillaries is known as the **peritubular capillary network**. Blood is collected from the peritubular capillary network by venules that conduct the blood into the renal vein.

Distribution of Nephrons in the Kidney's Cortex and Medulla

The kidneys are covered at their surface by a thick layer of connective tissue called the **renal capsule**. Internally, the tissues of the organ give rise to two distinct regions: the outer **renal cortex** and the inner **renal medulla** (Fig. 12-3). The medulla surrounds an open chamber called the **renal pelvis**. The collecting ducts of the nephrons empty their contents into the renal pelvis, and the urine is then withdrawn from the pelvis by the peristaltic actions of the ureters.

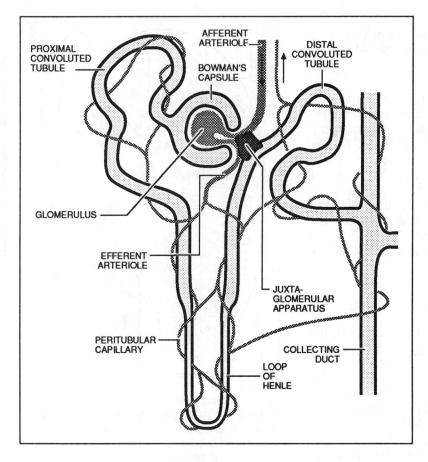

Figure 12-2
Structural organization of a nephron and its associated blood vessels.

183

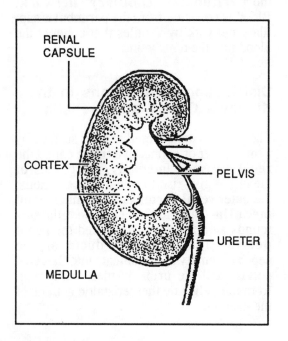

Figure 12-3
Internally, a kidney is divided into cortical, medullary, and pelvic regions (see text for explanation).

Figure 12-4 illustrates the manner in which the nephrons are arranged in the renal cortex and medulla. The Bowman's capsules and proximal and distal convoluted tubules of the nephrons are located in the renal cortex, whereas the loops of Henle and collecting ducts extend into the medulla. The arrangement resembles the manner in which the spokes of a wheel are arranged around the wheel's hub (i.e., the hub being analogous to the renal pelvis; see Fig. 12-4).

Mechanism of Nephron Action

The forces at play in the Bowman's capsule and glomerular capillary network result in the filtration of the blood. Two pressures characterize the blood in the glomerular capillaries: hydrostatic pressure (i.e., blood pressure) and colloid osmotic pressure (see Chapter 8 for a discussion of blood pressure and colloid osmotic pressure). As depicted in Figure 12-5, the blood pressure

(which is directed outward, against the capillary wall) is about 50 mm Hg and the colloid osmotic pressure (which is directed inward from the capillary wall) is about 30 mm Hg. There is, therefore, a net pressure gradient of 20 mm Hg (i.e., 50 mm Hg minus 30 mm Hg) acting to express fluid through the walls of the glomerular capillaries; this is called the **glomerular filtration pressure**. The glomerular filtration pressure is uniform over the entire length of the glomerular capillary network. This is in contrast with capillary beds of other tissues of the body in which the blood pressure falls precipitously. The maintenance of high blood pressure in the glomerular capillaries is due in part to the differential sizes of the afferent and efferent arterioles. The afferent arterioles, which conduct blood into the glomeruli, are wide vessels; in contrast, the efferent arterioles

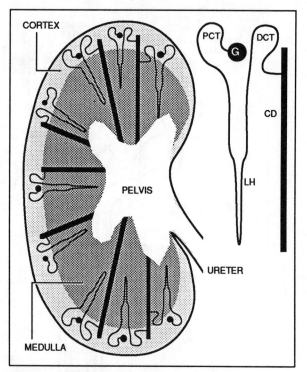

Figure 12-4
Within the kidneys, the nephrons are distributed in a radial pattern (like bicycle wheel spokes). G = glomerulus; PCT = proximal convoluted tubule; LH = loop of Henle, and DCT = distal convoluted tubule.

184

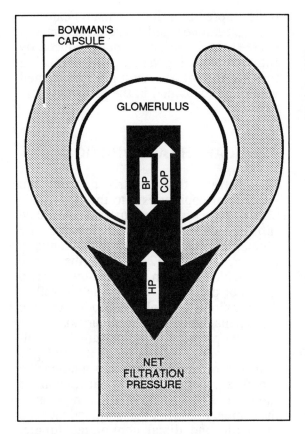

Figure 12-5
The blood pressure in the glomerular capillaries (50 mm Hg) exceeds the colloid osmotic pressure (30 mm Hg) by 20 mm Hg. This net pressure also dominates the hydrostatic pressure of the Bowman's capsules (10 mm Hg), so that there is an overall pressure gradient driving a filtrate of the blood into the Bowman's capsules.

are constricted. The narrowness of the singular opening through which the blood can exit the glomeruli acts to maintain high pressure within the glomeruli.

The Bowman's capsule also contains fluid (i.e., the filtrate of the blood) and this fluid exerts a hydrostatic pressure (see Chapter 8) of about 10 mm Hg. Because this pressure is directed against the walls of the Bowman's capsule, it opposes the filtration pressure of the glomerular capillaries. The difference between the glomerular filtration pressure (20 mm Hg) and the hydrostatic pressure of the Bowman's capsule is 10 mm Hg (i.e., 20 minus 10); it is this pressure difference that results in the movement of fluid from the glomerular capillaries into the Bowman's capsules.

The glomerular capillaries and Bowman's capsules of the nephrons are especially suited for filtration. The capillary walls and the walls of the Bowman's capsules are formed by a single, thin layer of epithelial cells and are, therefore, quite porous. The pores through which the filtration occurs are large enough to allow passage of substances with molecular weights up to about 30,000 daltons (see Chapter 1). Therefore, most of the small molecules of the blood are filterable (e.g., water, salts, sugars, amino acids, vitamins, etc.). The plasma proteins, however, exceed this size threshold and are not filtered (hence, no colloid osmotic pressure exists in the filtrate). Red and white blood cells are also far too large to be filtered from the

TABLE 12-1   COMPOSITION OF BLOOD PLASMA, FILTRATE, AND URINE

SUBSTANCE	AMOUNT IN PLASMA	AMOUNT IN FILTRATE	AMOUNT IN URINE
All Substances	550 c.c.	100 c.c.	< 2 c.c.
Water	500 c.c.	98 c.c.	< 2 c.c.
Protein	40.0 grams	0.00 grams	0.00 grams
Sugar (Glucose)	0.50 grams	0.09 grams	0.00 grams
Amino Acids	0.20 grams	0.04 grams	0.00 grams
Creatinine	0.005 grams	0.0009 grams	0.002 grams
Ammonia	0.000 grams	0.000 grams	0.005 grams

The quantities listed in each column are based on the passage of one liter of blood through the kidneys (i.e., from the renal arteries to the renal veins).

glomerular capillaries into the Bowman's capsules.

For each liter of blood that circulates through the kidneys, about 100 c.c. of filtrate is formed, and this filtrate contains a cross-section of all substances in the blood having molecular weights below 30,000 daltons. Because the hematocrit of a normal blood sample is about 0.45 (see Chapter 8), the 100 c.c. of filtrate formed each minute by the kidney tubules contains about 18.2% of all of the blood's small molecules (i.e., each liter of blood contains 550 c.c. of plasma [1000 − (.45 × 1000)], and 100 c.c. of filtrate is 18.2% of the 550 c.c.of filterable plasma).

Table 12-1 lists the quantities of some representative chemical substances that are present in the blood plasma and compares these with the amounts found in the filtrate of the Bowman's capsules and the urine. Although only six substances are listed in the table, hundreds of different chemical substances are present in normal blood plasma. For brevity, small ions such as Na^+, K^+, Cl^-, etc. are not listed in the table, but all of these ions are filterable and varying quantities are found in the filtrate and in the urine. The six substances that are listed in the table are representative of major chemical subgroups. Water is included because it is the major component to be filtered and is the medium in which all of the other substances are dissolved. Plasma protein is representative of non-filterable substances (i.e., their molecular weights are too great). Sugars and amino acids represent molecules that are small enough to be filtered but which are useful to the body and ought to be retained (not lost in the urine). Creatinine and ammonia represent true wastes that ought to be excreted *in toto* (i.e., all should be excreted from the body in the urine, not just the percentage that is filtered).

Especially striking are the values in the far right column of Table 12-1. These values show that *(a)* some substances filtered into the Bowman's capsules do not end up in the urine (e.g., glucose and amino acids), *(b)* some substances are present in greater amounts in the urine than are initially filtered into the Bowman's capsules (e.g., creatinine), and *(c)* some substances excreted in the urine are not produced by filtration of the blood (e.g., ammonia). The reasons for these seemingly unexpected values are explained in the following sections.

TUBULAR REABSORBTION

The filtrate that is driven from the glomerular capillaries into the Bowman's capsule passes into the proximal convoluted tubule, while the blood leaving the glomerulus enters the peritubular capillary network (see Fig. 12-2). Thus, the filtrate and the blood remain separated by only two layers of cells (i.e., the epithelial wall of the proximal convoluted tubule and the mesothelial lining of the capillaries).

Blood pressure in the peritubular capillary network falls from about 30 mm Hg in that part of the network surrounding the initial segment of the proximal convoluted tubule to 15-20 mm Hg in the regions of the capillary network that surround the loop of Henle and distal convoluted tubule. Whereas the blood's hydrostatic pressure falls, the blood's colloid osmotic pressure is maintained at about 30 mm Hg; this is because no plasma proteins are lost to the filtrate. Within the kidney tubule itself, the hydrostatic pressure remains at about 10 mm Hg. Considering the forces that are at play (Fig. 12-6), it should be apparent that there is now a net force directed from the filtrate back into the bloodstream.

> For example, if the blood pressure and colloid osmotic pressure in the peritubular capillary are 30 mm Hg (values that would exactly cancel one another), and if the hydrostatic pressure in the proximal convoluted tubule is 10 mm Hg, then there is a net pressure gradient of 10 mm Hg acting to push materials from the filtrate, across the walls of the tubule and capillary, into the bloodstream (i.e., 30 minus 30 minus 10).

The pressure gradient acting in the direction of the bloodstream causes water to reenter the blood of the peritubular capillary from the filtrate. The return of water to the bloodstream in this manner is called **obligatory water reabsorbtion** ("obligatory" because it is in response to pressure gradients over which the body does not exercise much control).

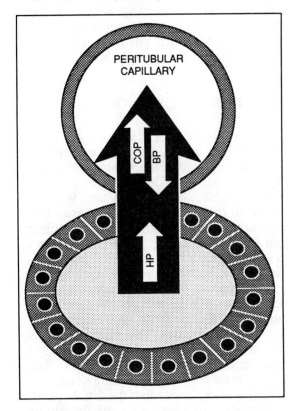

Figure 12-6
Because the blood pressure in the peritubular capillary is dominated by the blood's colloid osmotic pressure and the tubule's hydrostatic pressure, water is reabsorbed into the blood from the proximal convoluted tubule and loop of Henle. This is known as *obligatory water reabsorbtion.*

A number of other substances that were filtered into the Bowman's capsules are returned to the blood of the peritubular capillaries. Included here are the sugars and amino acids that were lost to the filtrate. Sugars and amino acids are reabsorbed by **active transport**; that is, enzymes in the membranes of the cells that line the proximal convoluted tubule actively pump sugars and amino acids from the lumen of the tubule into the bloodstream. Therefore, under normal circumstances, no sugars or amino acids are lost from the body in the urine. (In the condition known as **diabetes mellitus** [discussed in detail in Chapter 13], the blood glucose level rises far above its normal value, and the capacity for complete reabsorption is exceeded. This results in the appearance of glucose in the urine.)

Positive ions (especially Na^+) that were filtered from the blood of the glomeruli are also actively transported back into the blood of the peritubular capillary. In order to preserve the electrical neutrality of the filtrate and the blood, chloride ions passively accompany the reabsorbtion of the Na^+. The transport of sugars, amino acids, and ions back into the bloodstream is accompanied by the return of water. Indeed, by the time the filtrate reaches the end of the proximal convoluted tubule, about 65% of its water content has been drawn back into the blood by the combined effects of the pressure gradient and active transport (Fig. 12-7).

Reabsorbtion of salts and water continues as the filtrate passes through the loops of Henle. Sodium ions are actively transported back into the blood and are accompanied by the passive movements of chloride ions. The increasing tonicity of the blood and the falling tonicity of the filtrate that result from the movements of these ions cause water to return to the blood by osmosis. By the time the filtrate reaches the distal convoluted tubule, about 85% of all of the water originally lost from the blood by filtration has been reabsorbed.

In the distal convoluted tubule, chloride ions are actively transported back into the bloodstream (Fig. 12-7). The active transport of the Cl^- is accompanied by the passive return of Na^+ to the blood. No water is exchanged between the distal convoluted tubule and the blood of the peritubular capillary; indeed, the wall of the distal convoluted tubule is impermeable to water.

As a result of the reabsorbtion that takes place from the nephron into the peritubular capillaries, the filtrate that reaches the collecting ducts is considerably reduced in volume and is also very dilute. In the collecting ducts, the filtrate is concentrated by the further removal of water. The reabsorbtion of water from the collecting ducts into the bloodstream is subject to hormonal regulation and is called **facultative water reabsorbtion**. The hormone controlling this process is **antidiuretic hormone** (abbreviated ADH and also known as **vasopressin**; Fig. 12-7). ADH is produced by the **pituitary gland** at the base of the brain (see Chapter 13). When the amount of ADH released into the blood by the pituitary increases, there is increased facultative water reabsorbtion and the urine that is excreted from the body is more concentrated. When the amount of ADH released into the blood by the pituitary decreases, there is decreased facultative water reabsorbtion and the urine that is excreted from the body is more dilute.

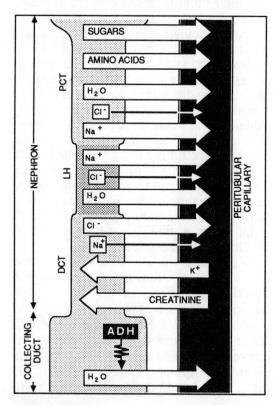

Figure 12-7
Tubular reabsorbtion and tubular secretion. Large arrows represent movements involving active transport; small arrows depict passive movements.

From the preceding discussion, it is clear that the composition of the glomerular filtrate is dramatically altered by tubular reabsorbtion as sugars, amino acids, sodium ions, and other substances of value are returned to the bloodstream. Not reabsorbed (or only minimally reabsorbed) are the metabolic wastes present in the filtrate. These wastes include creatinine, urea, and uric acid which are nitrogen-containing waste products of tissue metabolism.

TUBULAR   SECRETION

Since only 100 c.c. of filtrate are produced for each liter of blood that passes through the kidneys, most of the waste products in the blood are not filtered into the Bowman's capsules. Instead, they pass with the blood from the glomerular capillaries into the peritubular capillaries. It is now clear that a good part of this waste is transferred into the nephron from the peritubular capillaries. That is, the molecules move in a direction opposite to that of tubular reabsorbtion by crossing the walls of the peritubular capillaries and nephron and entering the lumen of the tubules. This phenomenon is called **tubular secretion** and is based upon active transport (Fig. 12-7). The active transport of creatinine from the peritubular capillaries into the distal convolute tubules explains why the creatinine concentration of the urine is so much greater than that of the original filtrate. Other substances are also subjected to tubular secretion, including potassium ions and hydrogen ions. The secretion of H+ into the filtrate explains why urine is so much more acidic than the blood plasma.

Although there is little or no ammonia present in the blood, urine does contain small amounts of ammonia (see Table 12-1); this warrants some explanation. Ammonia is a waste product of the breakdown of nitrogen-containing materials in the body (e.g., proteins, nucleic acids, etc.). Because most tissues are especially sensitive to ammonia and large quantities of this substance are toxic, waste ammonia is chemically modified prior to its release from a tissue. Tissue cells combine their ammonia with carbon dioxide (also a waste), thereby forming *urea*. The urea, which is considerably less toxic than ammonia, is then released into the blood along with other wastes. Where then does the ammonia present in urine come from? The cells that form the walls of the nephrons also produce ammonia as a waste product of their metabolism. These cells, however, can empty their waste directly into the glomerular filtrate (instead of the bloodstream). Consequently, the cells of the nephron's walls do not convert ammonia to urea; they release their ammonia directly into the filtrate, the ammonia ending up in the urine.

THE JUXTAGLOMERULAR APPARATUS

The wall of the distal convoluted tubule fuses with the wall of the afferent arteriole just before the arteriole's entry into the Bowman's capsule; this region of fusion is called the **juxtaglomerular apparatus** (Figs. 12-2 and 12-8). The juxtaglomerular apparatus plays an important role in the regulation of (1) blood pressure, (2) filtration into the Bowman's capsules, and (3) retention or excretion of sodium ions.

When the blood pressure in the systemic circulation falls, there is a corresponding reduction in the glomerular filtration rate (recall that it is the blood pressure that provides the force of filtration). Under these circumstances, the volume of filtrate proceeding through the various segments of the nephron is reduced. The reduced flow of filtrate in the distal convoluted tubule is sensed by the **macula densa** cells (Fig. 12-8) which then cause the **granular cells** to release the enzyme **renin** into the bloodstream. Renin then acts on a plasma protein called **angiotensinogen**, converting this protein to **angiotensin I**. Angiotensin I is then converted in the lungs to

Figure 12-8
The walls of the distal convoluted tubule fuse with the wall of the afferent arteriole, just prior to the arteriole's entry into the Bowman's capsule. This region of fusion, called the **juxtaglomerular apparatus**, plays an important role in regulating blood pressure, glomerular filtration, and salt balance. See text for details.

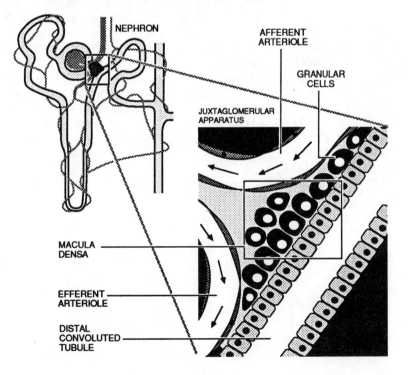

angiotensin II, a powerful vasoconstrictor. The generalized vasoconstriction that is promoted by angiotensin II acts to raise the blood pressure. Angiotensin II also stimulates the pituitary gland to release additional quantities of antidiuretic hormone (see earlier), thereby promoting greater water reabsorbtion from the kidney's collecting ducts. The increased return of water to the bloodstream helps to raise the blood pressure. Among the blood vessels constricted through the effects of angiotensin II are the efferent arterioles. Constriction of the efferent arterioles raises the blood pressure in the glomerular capillaries, and this promotes greater filtration of wastes into the Bowman's capsules.

Angiotensin II also acts on the body's adrenal glands, stimulating the release of the hormone **aldosterone**. This hormone acts on the kidneys, promoting the reabsorbtion of sodium ions into the bloodstream from the glomerular filtrate. Sodium reentering the blood is accompanied by water, thereby further elevating the blood pressure.

MICTURITION

Urine passes from the collecting ducts into the pelvis of the kidney, where it is then swept downward into the urinary bladder by the peristaltic waves of the ureters (Fig. 12-1). Emptying of the urinary bladder is facilitated by the parasympathetic division of the autonomic nervous system and is referred to as **micturition**. The thick, muscular walls of the urinary bladder are quite elastic, with the result that about 400 c.c. of urine can be accepted by the bladder before appreciable tension is developed in the bladder wall. During the gradual filling of the bladder, the wall is maintained in a relaxed state and the inner and outer **sphincter muscles** that encircle the urethra remain contracted.

When the volume of urine in the bladder exceeds 400 c.c., rapidly developing tension stimulates stretch receptors in the bladder's walls and results in the flow of sensory signals to the spinal cord. These sensory signals trigger the flow of motor impulses from the autonomic nervous system's parasympathetic division in the sacral region of the spinal cord. The motor impulses cause the bladder's walls to contract and the urethral sphincters to relax. The result is the expulsion of urine through the urethra to the surface of the body. Contractions of the bladder wall continue until the entire volume of urine has been voided. Bladder wall contraction simultaneously squeezes shut the opening of the ureters at the rear of the bladder, temporarily halting the flow of urine from the kidneys into the bladder. Once the bladder is emptied, parasympathetic impulses are halted, and the bladder begins to fill again.

In young children, the reflexive filling and emptying of the urinary bladder is automatic. With increasing age, and through training, it soon becomes possible to override the spinal reflexes so that emptying of the bladder can be delayed until it is socially convenient. In the same regard, it is also possible to initiate the voiding of urine even when the bladder is only partly filled (for example, when asked to provide a urine sample for clinical analysis).

CHAPTER 13

THE ENDOCRINE SYSTEM

In order for the body's functions and actions to be fully integrated, one organ (or tissue) must be able to communicate with another. The most rapid way of achieving such communication involves the nervous system. As you learned in Chapter 5, nerve fibers rapidly conduct impulses from one body part to another, ultimately bringing about physiological and/or mechanical changes.

Inter-organ communication may also be achieved in other ways. For example, certain tissues secrete "chemical messengers" into the bloodstream, and these are then carried through the circulatory system to target organs. Upon reaching the target organ, the chemical messages bring about physiological changes. Chemical messengers circulating in the blood are more appropriately called **hormones**, and the tissues that release hormones are called **endocrine glands**. Collectively, the endocrine glands comprise the body's **endocrine system**.

ENDOCRINE VS. EXOCRINE GLANDS

Before we proceed any further, it is important to point out that there are two different classes of glands in the body: **endocrine glands** and **exocrine glands** (Fig. 13-1). Exocrine glands deliver their secretions into small ducts which then convey the secretion onto an external or internal body surface, such as the skin or the lumen of the digestive tract (e.g, oral cavity, stom-

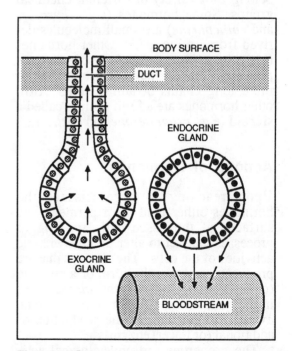

Figure 13-1
Comparison of exocrine and endocrine glands (see text for explanation).

ach, intestine). As a result, the secretion has an effect at (or near) the site of the gland. For example, water secreted by the sweat glands in the skin of the face cools the face; digestive enzymes secreted by the salivary glands of the mouth serve to digest food in the mouth; and so on. In contrast, the secretions of the endocrine glands may have effects at great distances from the location of the gland. This is because the glands' secretions enter the bloodstream and are carried throughout the body via the circulatory system. There is an added implication to secretion into the bloodstream; namely, that the secretions may affect several different and widely-separated body parts. Because they lack ducts, the endocrine glands are sometimes referred to as the body's *ductless glands*.

HORMONES

Chemistry of Hormones

Hormones secreted by the endocrine glands belong to a variety of different chemical groups. Some hormones (e.g., *thyroxin* and *epinephrine*) are small molecules derived from **amino acids**. Other hormones (e.g., *insulin* and *erythropoietin*) are **proteins**; that is, they consists of one or more long chains of amino acids. Still other hormones are a form of lipid called a **steroid** (e.g., *progesterone* and *cortisol*).

Actions of Hormones

Upon reaching a target tissue, the hormones either bind to or permeate the surfaces of the tissue cells and trigger processes that act to alter the biosynthetic activities of the cells. The changes that are produced may take the form of activating (or deactivating) enzymes that *already* exist in the cell, or the effect may be on *gene expression*, resulting in the production of additional (or fewer) enzymes.

The properties and physiological activities of cells are determined to the largest extent by the kinds of proteins that the cell possesses, especially the cell's enzymes. These proteins are the cell's functional machinery and are responsible for the specific actions that characterize a cell's (or tissue's) behavior. By affecting the cell's proteins, hormones can change the behavior of cells in several different ways. For the most part, the hormones bring about either (1) a change in the *nature* of these catalytic proteins, or (2) a change in the *amount* of enzyme that is present. To change the amount of enzyme, it is necessary to alter the rate of enzyme synthesis, and this requires that the hormone somehow alter the expression of the cell's genetic material in the cell nucleus.

To fully appreciate the effects of hormones, it is necessary to understand the mechanism by which a cell's enzymes (and other proteins) are synthesized. The process is illustrated in simplified form in Figure 13-2. The typical human cell contains thousands of different proteins. Among the things that make one protein different from another are *(a)* the numbers of amino acids in the protein, *(b)* the kinds of amino acids that are present (recall that there are about 20 different amino acids that regularly occur in human proteins), and *(c)* the order in which the amino acids occur in the polypeptide chains that comprise the protein. For every human protein, these three characteristics are encoded in one or more genes present in the cell nucleus.

The genetic material in the cell nucleus is composed of DNA (i.e., deoxyribonucleic acid), the DNA taking the form of two intertwined polynucleotide chains that form a double helix (see Chapter 2). For each gene that encodes a protein, it is the sequence of nitrogenous bases (one nitrogenous base in each nucleotide) along the polynucleotides that dictates the order of amino acids in the encoded protein. Whereas the genes that encode the cell's proteins are in the cell nucleus, the synthesis of proteins takes place in the extranuclear cytoplasm.

To produce a particular protein, the genes that encode the protein's structure must first be *transcribed*. During **transcription**, which takes place within the cell nucleus, the order of nitrogenous bases

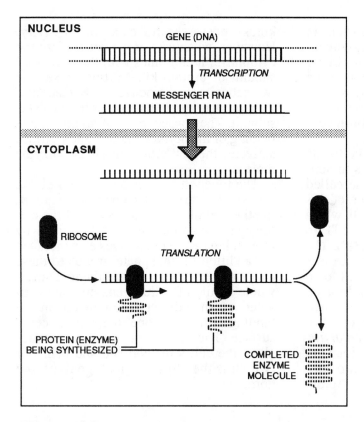

Figure 13-2
Transcription and translation of protein-encoding genes (see text for explanation).

along one of the gene's two polynucleotides is used as a *template* for the synthesis of a related polynucleotide called **messenger ribonucleic acid** (or mRNA). In mRNA, the order of nitrogenous bases is *complementary* to that of the transcribed polynucleotide.

Messenger RNA passes from the cell nucleus into the cytoplasm where it combines with **ribosomes**. The ribosomes, together with a number of other factors contributed by the cytoplasm, *translate* the mRNA's nitrogenous base sequence into a particular amino acid sequence. This is done by "reading" the message as a sequence of "codons," each codon consisting of three successive nucleotides. Consequently, produced in the cytoplasm by the **translation** process are proteins whose amino acid lengths and amino acid sequences reflect the various mRNAs produced by gene transcription. Each mRNA

molecule may be translated hundreds (or even thousands) of times, so that many copies of the final protein are produced from a single gene. (For a more detailed description of this mechanism consult any college-level introductory biology textbook.)

Protein hormones and steroid hormones differ in the manner in which they alter cellular and tissue activity. Therefore, we will consider the actions of the protein and steroid hormones separately.

Specific Actions of Protein Hormones. The surfaces of cells (i.e., plasma membranes) contain receptor molecules that bind chemical substances circulating in the bloodstream. The receptors are membrane proteins, and among the substances that they bind are circulating protein hormones (Fig. 13-3). The binding of the hormone by the receptor triggers chemical reactions in the cell's plasma membrane that result in the production of a "second messenger" (the "first messenger" is the hormone molecule itself). The most common of the second messengers produced by a hormone's target cells is a small compound called **cyclic adenosine monophosphate** (**cyclic AMP** or **cAMP**).

The attachment of the hormone to the plasma membrane receptor serves to activate an enzyme in the membrane called **adenylate cyclase**. The active site of this enzyme faces the interior of the cell, where the enzyme converts cytoplasmic ATP to cAMP and pyrophosphate (Fig. 13-3). The cAMP produced by the reaction diffuses into the cytoplasm where it attaches to an inactive enzyme called **protein kinase**. Protein kinase molecules consist of two subunits (i.e., two polypeptide chains). One of the subunits is called the *catalytic* subunit, whereas the other subunit is an *inhibitory* subunit. So long as the two subunits are attached to one another, protein kinase possesses no catalytic activity. However, when cAMP is released into the cytoplasm, it combines with the inhibitory subunit of protein kinase; this causes the two subunits to dissociate. Dissociation of the inhibitory subunit activates the catalytic subunit. The consequence of all this is that binding of the hormone at the cell's surface activates the cell's protein kinase enzyme molecules.

The function of activated protein kinase is to enzymatically attach phosphate groups to other cellular proteins. Some of the cellular proteins that are phosphorylated by protein kinase are themselves enzymes, and the addition of phosphate groups to these enzymes serves to activate them. Thus, a whole family of cellular enzymes are activated as an indirect result of the initial binding of the hormone to the target cell's surface. The activation of these enzymes alters the cell's metabolism and behavior (which is the physiological goal of the hormone).

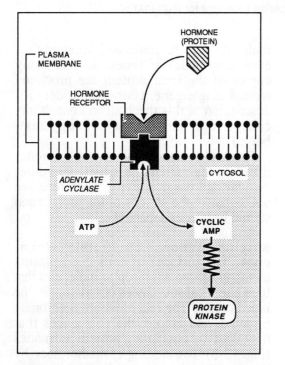

Figure 13-3
Actions of protein hormones on their target cells (see text for explanation).

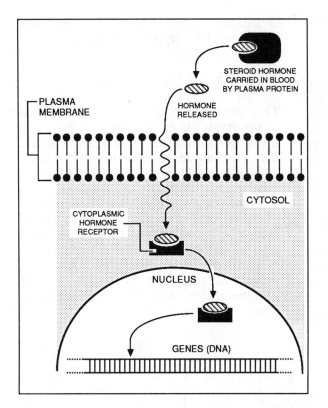

Figure 13-4
Action of steroid hormones on their target cells (see text for explanation).

The cyclic AMP that is produced in response to hormone binding at the cell surface is degraded within the target cell by an enzyme called **phosphodiesterase**. Therefore, the sustained alteration of a cell's metabolism requires the continuous production of fresh cAMP at the plasma membrane. This, in turn, requires that additional hormone be bound by the plasma membrane's hormone receptors.

In addition to cAMP, several other second messengers have been discovered. Included among these is **cyclic guanosine monophosphate** (or **cGMP**, a compound that is chemically quite similar to cAMP).

Specific Actions of Steroid Hormones. Like protein hormones, steroid hormones secreted into the blood by endocrine glands circulate in the bloodstream until they reach their target tissue. Because most steroids do not dissolve readily in water and are, there-fore, poorly soluble in blood plasma, they are specifically transported within the plasma by carrier proteins. Upon reaching the target cells, the steroid hormone separates from the carrier protein, permeates the target cell's plasma membrane, and enters the cytoplasm (Fig. 13-4). Here the steroid hormone combines with a cytoplasmic receptor molecule to form a **hormone-receptor complex**. The complex then enters the nucleus of the cell where it attaches to specific genes. The effect on the gene(s) to which the hormone-receptor complex binds is to "turn the gene on;" that is, gene transcription begins in genes that were not being transcribed prior to binding the hormone-receptor complex. As a result, new mRNA is produced, which passes into the cytoplasm from the nucleus and is translated into protein. The new proteins appearing in the cell (which usually are enzymes) bring about a change in the cell's metabolism and behavior.

195

A SURVEY OF THE MAJOR ENDOCRINE GLANDS

What follows is a survey of the body's major endocrine glands: (1) the **pituitary** gland, (2) the **thyroid** gland, (3) the **parathyroid** glands, (4) the **pancreas**, (5) the **adrenal** glands, (6) the **kidneys**, and (7) the reproductive tissues (i.e., the **ovaries** in females and the **testes** in males). For each of these glands, we will examine the hormones that they secrete and the effects of the hormones on their target tissues. Abnormalities in which the glands produce either too much or too little hormone will also be considered. Excessive production of hormone is called *hyperproduction* or *hyperactivity* (i.e., "hyper" = "over"); when too little hormone is se-

creted, this is called *hypoproduction* or *hypoactivity* (i.e., "hypo" = "under").

THE PITUITARY GLAND

The pituitary gland (also called the **hypophysis**) is sometimes referred to as the body's "master" endocrine gland because it influences so many body tissues and also regulates the behavior of many other glands. With this in mind, it may not be surprising to learn that the pituitary gland is located at the base of the brain (Fig. 13-5). The undersurface of the brain rests on a bony ledge formed by the skull. A small pocket in this ledge houses and protects the pituitary gland, which descends into the pocket from the overlying brain tissue.

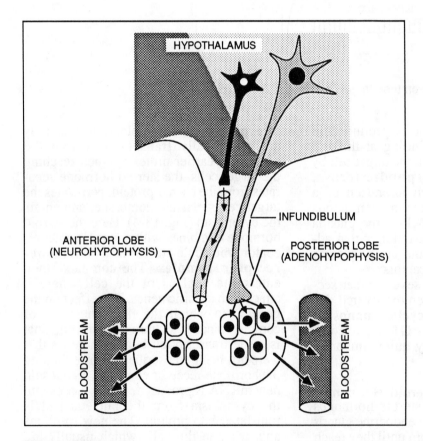

Figure 13-5
The anterior and posterior lobes of the pituitary gland.

Anterior Lobe (Adenohypophysis)

The pituitary gland is divided into front and rear halves, each half, or *lobe*, secreting separate families of hormones into the bloodstream. The front or **anterior lobe** (which is also called the **adenohypophysis**) secretes six hormones: (1) **human growth hormone** (abbreviated **HGH**), which is also known as **somatotrophic hormone** (abbreviated **STH**), (2) **thyrotropin** or **thyroid-stimulating hormone (TSH)**, (3) **adrenocorticotrophic hormone (ACTH)**, (4) **prolactin** (or **lactogenic hormone**), and the two *gonadotrophic hormones* (5) **follicle-stimulating hormone (FSH)**, and (6) **luteinizing hormone (LH;** also called **interstitial cell-stimulating hormone [ICSH])**.

Role of the Hypothalamus

The pituitary gland lies just below a region of the brain called the hypothalamus, which plays a major role in regulating the actions of the gland. Systemic blood that has circulated through the capillary networks of the hypothalamus does not return directly to the venous circulation; instead, this blood enters a *portal system* that carries it from the hypothalamus into the adenohypophysis. As illustrated in Figure 13-5, neurosecretory cells in the hypothalamus release a variety of hormones into the portal system. When these hormones reach the adenohypophysis, they cause the cells of the adenohypophysis to secrete their hormones. The hormones released into the portal blood by the hypothalamus include **growth hormone-releasing hormone, thyrotropin-releasing hormone, corticotropin-releasing hormone, prolactin-releasing hormone**, and **gonadotropin-releasing hormone**. Note that the names of the hypothalamic hormones reflect the names of the adenohypophyseal hormones whose secretion they promote. For example, growth-hormone-releasing hormone promotes the secretion of growth hormone, thyrotropin-releasing hormone promotes the secretion of thyrotropin, and so on.

Human Growth Hormone. Human growth hormone (HGH) influences tissue growth, and is especially important during the first 15-20 years of life when most of the body growth takes place. The hormone's influences are especially marked on the growth of the skeleton. Hypoproduction of HGH during the normal growing years results in reduced growth. In its extreme form, hypoproduction of HGH results in **pituitary dwarfism**, a condition in which the affected person retains childlike size and proportions. Although physical growth is diminished, mental development is entirely normal.

Hypersecretion of HGH produces varied effects, depending upon when in life the excessive production and secretion of the hormone occurs. Overproduction of HGH during the normal growing years results in a condition known as **gigantism** and is characterized by the affected person reaching extraordinary size (e.g., people who are 7 or 8 feet tall). When hyperactivity occurs after the normal growing years have ended, there is a different effect. Most bone growth occurs at the ends of the bones (i.e., at the **epiphyses**), and when skeletal growth is concluded, the epiphyses seal over. When there is excessive production of HGH after the epiphyses seal over, the bones grow wider and become distended. The condition is known as **acromegaly** and is characterized by distortion of the facial bones and the bones of the arms and hands and the legs and feet.

Thyroid-Stimulating Hormone. Thyroid-stimulating hormone (abbreviated TSH and also known as **thyrotropin**) acts upon the thyroid gland (see below), causing the gland to produce and secrete the thyroid hormones.

Adrenocorticotrophic Hormone. As we will see later, the tissues of the adrenal glands are divided into an outer region called the **adrenal cortex** and an inner re-

197

gion called the **adrenal medulla**. The adrenal cortex and the adrenal medulla produce and secrete different sets of hormones. Whereas pituitary secretions have no effect on the medulla of the adrenal glands, adrenocorticotrophic hormone (ACTH) released by the pituitary's anterior lobe acts to stimulate hormone production and secretion by the adrenal cortex.

Prolactin. Although prolactin is secreted by the anterior lobe of the pituitary gland in both males and females, it is only in females that the hormone's function is understood. In females, prolactin acts to promote the production and release of milk from the mammary glands after the birth of a baby. Note that in this case, an endocrine gland (i.e., the pituitary) is stimulating secretion by exocrine glands (i.e., the mammary glands).

Follicle-Stimulating Hormone. Follicle-stimulating hormone (FSH) is one of the two *gonadotrophic hormones* secreted by the adenohypophysis. In females, FSH promotes the development of follicles (the chamber-like structures that contain and protect developing egg cells; see later and Chapter 14) in the ovaries. In males, FSH promotes the production of sperm cells in the testes.

Luteinizing Hormone. Luteinizing hormone (LH) is the other gonadotrophic hormone produced by the adenohypophysis. In females, LH plays a role in inducing the monthly release of mature egg cells from the ovaries into the **Fallopian tubes** (again, see later and Chapter 14). LH also promotes the conversion of ruptured ovarian follicles into **corpora lutea**. In males, LH acts on the **interstitial cells of Leydig** in the testes; these are the cells that produce and secrete the male sex hormones (see below). LH is also called interstitial cell-stimulating hormone (ICSH).

Posterior Lobe (Neurohypophysis)

The cells that make up the posterior lobe of the pituitary gland do not actually synthesize the hormones that they secrete. Instead, the hormones are made in the overlying hypothalamus and are carried by the axons of hypothalamic neurons down into the pituitary's posterior lobe. The cells of the posterior lobe store the hormone molecules until instructed to secrete them. The hormones are secreted in response to nerve signals received from the hypothalamus (Fig. 13-5).

Two hormones are secreted by the pituitary's posterior lobe: **antidiuretic hormone (ADH**; also known as **vasopressin**) and **oxytocin**. Both neurohypophyseal hormones are small peptides consisting of a sequence of nine amino acids. Only two of the nine amino acid positions distinguish the hormones chemically. From a physiological point of view, however, ADH and oxytocin are quite distinct.

Antidiuretic Hormone. ADH was discussed in connection with kidney function in Chapter 12. This hormone acts on the kidneys' collecting ducts, promoting the return of water to the bloodstream during **facultative water reabsorbtion**. The greater the amount of ADH released into the bloodstream from the pituitary, the greater is the amount of facultative water reabsorbtion, and the more concentrated is the excreted urine. In contrast, when the amount of ADH released into the blood falls, there is less facultative water reabsorbtion and the excreted urine is greater in volume and more dilute. By influencing the amount of water that is returned to the blood from the kidney tubules, ADH also affects blood pressure (hence the alternate name, *vasopressin*).

Oxytocin. Although oxytocin is produced in both males and females, the hormone has no known function in males. In females, oxytocin has two effects. The first effect occurs during *parturition* (childbirth); at this time, oxytocin released into the blood acts to stimulate the forcible contractions of the smooth musculature of the wall of the uterus, thereby helping to expel the baby

198

from the uterus. Oxytocin also causes constriction of the uterine blood vessels reducing the loss of blood. After childbirth, the continued release of oxytocin promotes contraction of smooth muscle tissue around the ducts of the mammary glands; this promotes **lactation** (i.e., the release of milk from the glands).

THYROID GLAND

The thyroid gland consists of two lobes of tissue seated on the right and left sides of the trachea, just below the larynx (Fig. 13-6). The two lobes of the thyroid are joined on the anterior (i.e., front) surface by a narrow strip of thyroid tissue called the **isthmus**. The major secretions of this gland are **thyroxine**, **triiodothyronine**, and **calcitonin**.

Thyroxine and Triiodothyronine

Thyroxine and triiodothyronine are unusual substances in that they contain the element **iodine**. As its name suggests, each molecule of triiodothyronine contains three atoms of iodine; thyroxine, however, contains four iodine atoms (and is also known as **tetraiodothyronine**). For simplicity, the two hormones are usually referred to as T_3 (triiodothyronine) and T_4 (tetraiodothyronine).

The hormone-producing cells of the thyroid are arranged to form hollow spheres called **follicles** (Fig. 13-7). At the center of each follicle, there is a chamber containing a colloidal suspension of the protein **thyroglobulin**. Dietary iodine absorbed into the bloodstream is removed by the follicle cells as the blood flows through the thyroid tissue. The iodine is then transferred to the follicle chambers and temporarily stored in association with the thyroglobulin. Individual atoms of thyroglobulin-bound iodine are sequentially added to molecules of the amino acid **tyrosine**, first forming **monoiodotyrosine** and then **diodotyrosine**. These molecules are then removed from the colloid by the follicle cells and used to synthesize triiodothyronine and thyroxine.

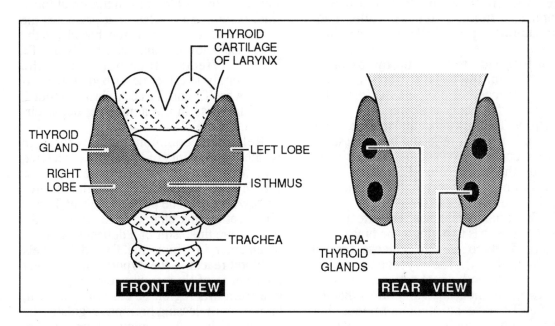

Figure 13-6
The thyroid and parathyroid glands.

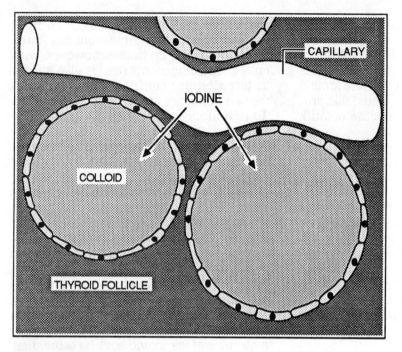

Figure 13-7
Iodine removed from the bloodstream is incorporated into thyroglobulin within the thyroid follicles.

T3 and T4 are secreted into the bloodstream where they combine with plasma proteins, forming a complex called **protein-bound iodine** (PBI). Upon reaching the target tissue, the hormones are released from PBI and are taken up by the target cells. Within the cells, the hormones are bound by receptor proteins that enter the nucleus and affect gene expression.

T3 and T4 bring about an increase in cell metabolism that is reflected by increases in oxygen consumption and the production of heat by the target cells. The effect is believed to be due to stimulation of the synthesis of electron carrier proteins of the mitochondrial membranes. The effect of T3 and T4 in liver is to promote the breakdown of stored glycogen into glucose (called **glycogenolysis**). The glycogenolytic effect is due to both direct action of the hormones and to the stimulatory effect that the T3 and T4 have on other hormones that promote glycogenolysis (e.g., *epinephrine*; see later).

T3 and T4 also promote the uptake of glucose into the blood from the small intestine. *Insulin*, a hormone that promotes the removal of sugar from the blood by the liver (see below), is inhibited by T3 and T4. These effects, together with the glycogenolytic effect described above, explain why T3 and T4 act to bring about a rapid increase in the amount of sugar circulating in the blood.

Relatively common are disorders in which the amounts of T3 and T4 secreted into the bloodstream are either too low (i.e., *hypothyroidism*) or too high (i.e., *hyperthyroidism*). The proper levels of T3 and T4 secretion that characterize normal individuals are the result of a delicate feedback mechanism in which the T3 and T4 levels of blood reaching the hypothalamus affect the release of thyrotropin-releasing hormone from this region of the brain. In turn, thyrotropin-releasing hormone regulates the amount of thyroid-stimulating hormone that is released by the adenohypophysis.

Hypothyroidism

Hypothyroidism is an abnormality in which the thyroid gland fails to secrete normal amounts of T3 and T4. When hypothyroidism is the result of a dietary iodine insufficiency, one of the symptoms is the abnormal growth of the thyroid gland, producing a massive enlargement of the throat known as an **endemic goiter**. The goiter is the result of a disturbance of the feedback loop that controls the thyroid's size. Because the thyroid is not secreting enough T3 and T4, the T3 and T4 levels reaching the hypothalamus are abnormally low. In an effort to raise the T3 and T4 levels, the hypothalamus responds by releasing more thyrotropin-releasing hormone. This causes the anterior lobe of the pituitary to release more thyroid-stimulating hormone. The elevated level of thyroid-stimulating hormone causes excessive growth of the thyroid gland.

Hypothyroidism in a baby is difficult to detect because during pregnancy T3 and T4 are transferred across the placental membranes from the mother's blood to the baby's blood. As a result, a hypothyroid baby appears normal. However, hypothyroidism during infancy leads to a condition called **cretinism**, characterized by retarded physical and mental growth, abnormal bone development, low body temperature, and a general lethargy. If diagnosed sufficiently early, these effects can be reversed by administering T3 and T4.

Hypothyroidism in adults is characterized by a condition known as **myxedema** in which an excessive amount of tissue fluid fills the spaces between tissue cells and the bloodstream. The skin, especially the facial skin, appears abnormally puffy. Other symptoms include low body temperature, weight gain, and a general lethargy.

Hyperthyroidism

Hyperthyroidism is an abnormality in which the thyroid gland secretes abnormally high amounts of T3 and T4. The disease is rare in infants, but in adults is accompanied by high heart rate and blood pressure, a general irritability (i.e., excessive "nervous energy"), and loss of weight. Two especially obvious physical symptoms are **exophthalmus** and **toxic goiter** (also called **Grave's disease**). In instances of toxic goiter, it is the enlarged thyroid that is the source of the excessive amounts of T3 and T4 that are secreted. Toxic goiters usually are not as large as endemic goiters.

Calcitonin

Blood contains small amounts of calcium derived by absorbtion from the digestive tract or by decalcification of bone. Calcitonin, together with parathyroid hormone (see below), affects the level of calcium in the blood. Calcitonin also influences the amount of phosphate excreted from the body. The hormone is a short polypeptide.

PARATHYROID GLANDS

The parathyroid glands (Fig. 13-6) consist of four small masses of tissue buried in the rear surface of the thyroid gland (two nodules in each lobe of the thyroid). The major hormonal secretion of this gland is a polypeptide called **parathormone** (also called **parathyroid hormone**, or **PTH**). Parathormone's principal effect is the regulation of the levels of calcium (and to a lesser extent, phosphate) that are circulating in the blood. An increase in the amount of parathormone being secreted into the bloodstream is followed by an increase in the blood's calcium (and phosphate) levels. Normally, the increase in blood calcium is the result of increased absorbtion through the digestive tract (the major source of calcium being milk). Since PTH increases the level of calcium in the blood, its effects are *antagonistic* to those of calcitonin.

PANCREAS

We considered the pancreas previously in

201

connection with the digestive system and learned then that the pancreas is a major source of digestive enzymes. In addition to its role as a digestive organ, the pancreas is an important endocrine gland. The pancreas produces two major hormones: **insulin** and **glucagon**. Both hormones are small proteins and are involved in the regulation of sugar metabolism in the body. Like calcitonin and parathormone, insulin and glucagon are antagonistic (i.e., they have opposite effects).

The pancreatic digestive enzymes and the pancreatic hormones are produced and secreted by different cells. The enzyme-producing cells, called **acinar cells**, make up most of the pancreatic tissue; this is *exocrine* tissue, since its secretory products exit the pancreas through ducts (which convey the enzymes toward the duodenum). The pancreas' endocrine tissue consists of clusters of cells scattered through the acinar tissue. The cell clusters, called **Islets of Langerhans** (Fig. 13-8), are surrounded by capillaries into which the hormonal secretions are emptied.

Insulin and glucagon are secreted by two different types of cells in the islets. The so-called **beta cells** (which make up about 70% of the cells in each islet) secrete insulin, whereas the **alpha cells** (approximately 30% of the islet cells) secrete glucagon.

Insulin acts to lower the amount of sugar circulating in the blood by promoting the uptake of sugar by the liver and by adipose (i.e., fat) tissue scattered through the body. Within the liver, insulin promotes the conversion of sugar to glycogen (glycogen is a high molecular weight polymer of glucose and is the principal form in which sugar is stored in the body; see Chapter 4). In adipose tissue, insulin acts to promote the conversion of sugar to fat. During exercise, insulin promotes the transfer of glucose into muscle cells.

In addition to its effects on sugar metabolism, insulin promotes the uptake of amino acids from the blood and the incorporation of these amino acids into cellular proteins. Consequently, together with human growth hormone, insulin is required for proper tissue growth.

Glucagon is antagonistic to insulin, promoting the breakdown of glycogen into glucose and the release of this sugar into

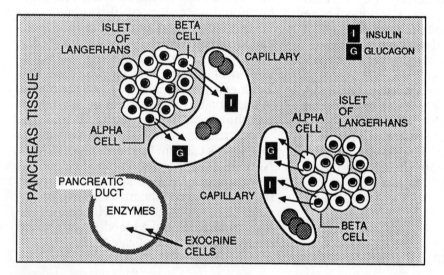

Figure13-8
Pancreas tissue consists principally of digestive enzyme-producing acinar cells.
Scattered through tissue are the *Islets of Langerhans*, whose alpha and
beta cells produce the hormones glucagon and insulin.

the blood. The pancreas secretes insulin and glucagon in response to the amount of sugar circulating in the bloodstream. Consequently, a delicate interplay between these two hormones ensures that the blood sugar level is neither too high nor too low.

Diabetes mellitus

There are two relatively common physiological abnormalities involving the secretion and actions of the pancreatic hormones; these are the genetically determined diseases called **diabetes mellitus type I** (about 10% of all cases) and **diabetes mellitus type II** (about 90% of all cases). In diabetes mellitus type I, there is an insulin deficiency (that is *hyposecretion* of insulin) due to malfunction of the beta cells of the islets of Langerhans. In diabetes mellitus type II, insulin secretion is approximately normal, but the target tissues do not respond with increased sugar uptake from the blood. The diseases' most obvious clinical symptoms are abnormally high plasma glucose levels following a meal (i.e., **hyperglycaemia**) and the appearance of sugar in the urine. The origin of these symptoms may be explained as follows.

In a normal person, the digestion of carbohydrate and the absorbtion of the resulting sugars begins to raise the blood sugar level. However, the rise in blood sugar is paralleled by increased secretion of insulin into the bloodstream. As a result, sugar uptake by the liver and adipose tissue is stimulated. Recalling the discussion of the digestive system presented in Chapter 11, sugars absorbed into the bloodstream following digestion of carbohydrate enter the **hepatic portal system**. The portal system diverts venous systemic blood through the narrow sinuses of the liver before the blood enters the inferior vena cava and returns to the right side of the heart. Sugar removed from the portal blood by the liver is stored as glycogen. Small amounts of the sugar are allowed to re-enter the blood as sugar is needed to fuel tissue metabolism between meals. Now, in diabetes mellitus types I and II, the amount of sugar removed from the blood (following digestion of carbohydrate and absorbtion of sugar) is diminished. Reduced sugar removal by the liver causes the blood sugar level to rise above normal. Recalling the discussion in Chapter 12 of the physiology of the kidneys, sugar molecules are sufficiently small to be filtered from the blood of the glomerular capillary networks into the Bowman's capsules. Whereas in a normal person the small amount of filtered sugar is not lost in the urine because the sugar molecules undergo tubular reabsorbtion, in a diabetic much of the filtered sugar is lost in the urine. This is because the amount of sugar that is filtered greatly exceeds that which can be reabsorbed into the bloodstream.

The loss of sugar in the urine is not, in itself, especially harmful. What is harmful is the variety of additional physiological consequences of elevated blood sugar levels. For example, the increased sugar level of the blood and the body dehydration that results from excessive urination increase the blood's viscosity, and this places an additional burden on the heart. Since insulin also promotes the uptake of amino acids and synthesis of protein by the body's tissues, protein metabolism is adversely affected. In diabetes mellitus type I, the insulin deficiency slows the rate of fat synthesis in adipose tissue. As a result, fatty acids may be released into the blood and converted in the liver to ketones. This leads to an elevated ketone level of the blood called **ketosis**.

Diabetes mellitus type I can be treated by administering exogenous insulin to the diabetic. Until quite recently, the only sources of insulin that were available were samples purified from the pancreases of pigs and cows. Although ovine and bovine insulins are chemically similar to human insulin, they are not identical and, as a result, they are not entirely effective. Now, however, it is possible to produce large amounts of human insulin using genetic engineering methodologies.

In some individuals suffering from diabetes mellitus type II, ingestion and digestion of carbohydrate may be followed by an excessive secretion of insulin. This, in turn, can cause too much sugar to be removed from the blood, thereby producing what is called **reactive hypoglycaemia** (i.e., too little blood sugar). When the amount of sugar circulating in the blood is too low, there usually are symptoms of hunger, weakness, impaired mental ability, and blurred vision.

THE ADRENALS

Seated on the upper surface of each kidney is an **adrenal gland** (Fig. 13-9). Each adrenal gland is divided structurally and functionally into two regions (Fig. 13-9). The outer portion of each gland is called the **adrenal cortex**; this region accounts for about 20% of the gland's mass. The central portion of each gland is called the **adrenal medulla** and accounts for the remaining 80% of the gland's mass.

Secretions of the Adrenal Cortex

In response to pituitary ACTH, the cortex of the adrenal glands secretes two classes of hormones known as **mineralocor-**

ticoids and **glucocorticoids**. All are steroids produced from cholesterol.

The principal mineralocorticoid is **aldosterone**; this hormone plays an important role in maintaining the body's salt balance. Aldosterone promotes the reabsorption into the bloodstream of Na+ filtered in the kidneys, while promoting the excretion of K+ through the kidneys. The main glucocorticoid is **cortisol**. This hormone has a variety of influences on metabolism. It promotes the synthesis of sugars from amino acids and promotes the storage of glycogen. The production and secretion of aldosterone and cortisol by the adrenals is regulated by ACTH released by the anterior lobe of the pituitary.

Although the ovaries in females and testes in males are the major sources of the sex hormones (see later), the adrenal cortex produces small quantities of androgens and estrogens.

Secretions of the Adrenal Medulla

In response to sympathetic nerve stimulation, the adrenal medulla secretes the **catecholamines epinephrine** (also called **adrenaline**) and **nor-epinephrine** (also

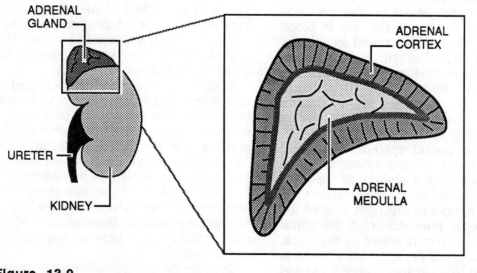

Figure 13-9
Organization of an adrenal gland.

204

called **nor-adrenaline**). Both of these hormones are **sympathomimetic**; that is, their effects mimic the sympathetic division of the autonomic nervous system. Epinephrine promotes an increase in blood sugar by mobilizing sugar stored in the liver as glycogen. Epinephrine also increases heart rate and blood pressure and acts as a vasodilator in skeletal muscle. Nor-epinephrine is a strong vasoconstrictor and, like epinephrine, raises the blood pressure.

THE KIDNEYS

The kidneys produce and secrete the hormone **erythropoietin**. This hormone regulates the production of red blood cells by the bone marrow. The discovery of this hormone is relatively recent and quite interesting. When an animal suffers a blood loss and becomes anemic, the bone marrow responds by increasing its output of new blood cells. That the increased erythropoietic activity is the consequence of hormonal stimulation was recognized for the first time when plasma from an anemic animal was injected into the bloodstream of a normal animal. The normal animal quickly developed a **polycythaemia**; that is, within a few days the numbers of circulating red blood cells increased to a level well above normal. It was suspected that something in the injected "anemic plasma" was the cause of the elevated erythropoietic activity. The plasma factor was eventually identified as a protein and named erythropoietin because of its stimulatory effects on erythropoiesis. Not long after the discovery of erythropoietin, the source of this hormone was identified as the kidneys.

Erythropoietin acts on the bone marrow's stem cells, causing their more rapid and more frequent development into red blood cell progenitors. Erythropoietin may also cause earlier release of developing red cells. For example, several days following a severe blood loss, it is not unusual to find larger numbers of reticulocytes or even normoblasts in the peripheral blood. The fact that the kidneys are the source of erythopoietin explains why people who suffer kidney losses and are placed on dialysis must also receive periodic transfusions of blood. In the absence of fully functional kidneys, the lack of erythropoietin secretion causes reduced bone marrow erythropoietic activity and this results in anemia.

THE REPRODUCTIVE ORGANS

The reproductive tissues (i.e., the **ovaries** in females and the **testes** in males) serve as endocrine glands in addition to their roles in the production of eggs and sperm.

The Ovaries

In order to appreciate the functions of the ovaries as endocrine glands, it is necessary to be familiar with the **ovarian cycle** and the role of the ovaries as the sources of egg cells. At birth, the ovaries of a female contain hundreds of thousands of egg cells (also called **ova** [singular = **ovum**] or **oocytes**). The egg cells are surrounded by a layer of non-reproductive cells called **follicle cells**. Collectively, each ovum and its surrrounding follicle cells comprise a **primary follicle** (Fig. 13-10; stage 1). The primary follicles remain dormant in the ovaries until the onset of puberty (12 to 15 years of age), when the ovarian cycle or **menstrual cycle** begins.

In the average female, each ovarian cycle lasts about 28 days, day number 1 being the first day of **menstruation** (i.e., loss of blood from the uterus). Beginning on day 1 of each cycle, about 20 primary follicles begin to undergo additional growth and development; that is, the number of follicle cells increases and the size of the egg cell increases. These maturing follicles are called **secondary follicles** (Fig. 13-10; stage 2). The growth of one secondary follicle dominates the others, and by about day 5, all of the secondary follicles except the dominant follicle cease development and begin to break down and disappear; these degenerating structures are called **atretic**

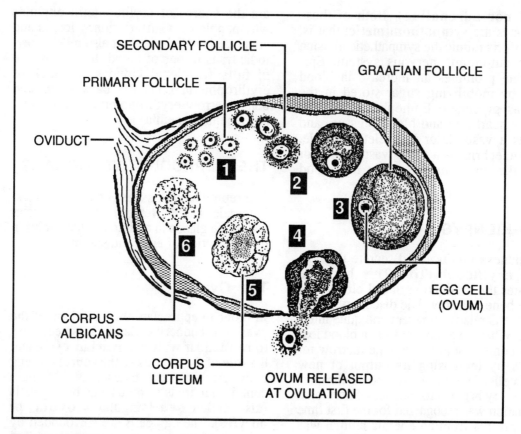

Figure 13-10
Various stages (numbered 1 through 6) of follicle development in an ovary during the ovarian cycle. See text for explanation.

follicles. The follicle that continues to grow is called a **Graafian follicle** and is characterized by several layers of follicle cells surrounding a central egg cell (Fig. 13-10; stage 3).

The development of the Graafian follicle is promoted by follicle-stimulating hormone that is being released by the anterior lobe of the pituitary gland. As the Graafian follicle grows, the follicle cells secrete **estrogen**, one of the two major ovarian hormones (actually estrogen is a mixture of three chemically similar steroids called **estradiol, estriol,** and **estrone**).

On day 14 of the menstrual cycle, the mature Graafian follicle fuses with the surface of the ovary, ruptures, and releases the ovum (together with some of the follicle cells) into the abdominal cavity (Fig. 13-10;

stage 4). This process, called **ovulation**, is believed to be triggered by a sudden rise in the level of luteinizing hormone released by the pituitary's anterior lobe, together with a more modest rise in follicle-stimulating hormone (Fig. 13-11).

After its release from the ovary, the ovum is guided into the **oviduct** (or **Fallopian tube**). The cells that line this tube are ciliated, and the beating of the cilia acts to convey the ovum toward the uterus. Meanwhile, the ruptured Graafian follicle is transformed into a **corpus luteum** (Fig. 13-10; stage 5). The cells of the corpus luteum (i.e., former Graafian follicle cells) continue the secretion of estrogen but in greater quantities. Also released by the corpus luteum are large amounts of a second ovarian hormone called **progesterone**. If

206

the egg cell released at ovulation is not fertilized and implanted in the wall of the uterus (the usual case), then beginning at about day 21 the corpus luteum begins to degenerate, and the secretion of estrogen and progesterone declines. The corpus luteum becomes a vestigial body called a **corpus albicans**. The changes in the levels of FSH, LH, estrogen, and progesterone that characterize the ovarian cycle are summarized in Figure 13-11.

The ovarian cycle is accompanied by cyclic changes in the lining of the uterus that prepare the uterus for reception and implantation of a fertilized egg. If no fertilized egg reaches the uterus, the lining disintegrates and is redeveloped in the next cycle. The breakdown of the uterine lining leads to a loss of blood from the uterus, or **menstruation**. As noted above, the first day of bleeding is taken as the first day of the 28-day ovarian cycle.

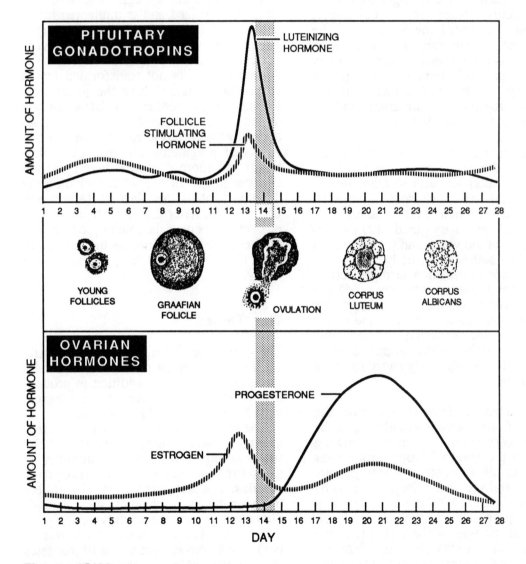

Figure 13-11
Changes in the levels of the gonadotropic hormones (i.e., follicle-stimulating hormone and luteinizing hormone) and ovarian hormones during the ovarian cycle.

Actions of the Ovarian Hormones

Estrogens. The estrogens have a variety of important influences on the body's functions and actions. From a metabolic point of view, the principal effects of estrogen are on protein synthesis and tissue growth. The most direct targets of estrogen action are the so-called *accessory reproductive organs*. At puberty, estrogen promotes the growth of of these accessory structures, and in the absence of estrogen, none of the changes associated with puberty transpire. Accordingly, estrogen stimulates the growth of the uterus, the development of the external genitalia, and the growth of the mammary glands. In adult females, estrogen is necessary for the maintenance of the accessory reproductive structures and the initiation and continuation of the menstrual cycle.

Estrogen also influences electrolyte balance (i.e., the excretion and retention of water and salts) and acts to maintain the female *secondary sex characteristics*, such as (1) the distribution of fat to the hips, abdomen, and mammary glands, (2) the elevated pitch of the voice, and (3) the pattern of hair distribution on the body.

At the time of ovulation, estrogen acts on the cervical glands so that they produce a secretion that favors the survival of sperm and facilitates sperm entry into the uterus from the vagina. Estrogen also increases the ciliary beating of the oviduct cells so that the ovum is efficiently swept from the ovary to the uterus.

Progesterone. Prior to ovulation, the amount of progesterone circulating in the blood is very small. However, following ovulation the level of progesterone rises rapidly, as this hormone and estrogen are produced and secreted in large amounts by the corpus luteum. Working in concert with estrogen, progesterone promotes the development of the wall of the uterus (i.e., the **endometrium**), so that the wall can accept and adequately nourish a fertilized egg and developing embryo (see also Chapter 14). The endometrium becomes thick and spongy and also highly vascular (i.e., rich in tiny blood vessels). If the egg cell released at ovulation is not fertilized, the secretion of progesterone declines. The fall in progesterone is followed by the progressive death of the tissues of the endometrium as the blood supply to the tissues is shut down. The dying (i.e., *necrotic*) tissue, along with small amounts of blood, is sloughed off the surface of the uterus and gives rise to the menstrual flow marking the onset of a new ovarian cycle.

In the event that the egg released at ovulation is fertilized and is implanted in the endometrial lining of the uterus, the corpus luteum remains intact in the ovary and continues to secrete progesterone (i.e., the corpus luteum is not transformed into a corpus albicans). During the pregnancy, follicle development and ovulation are also halted. Progesterone continues to be secreted by the corpus luteum for the first five weeks of pregnancy and promotes additional and more complex developmental changes in the uterine wall, including the development of the placenta (again, see Chapter 14). After this time, however, the placenta takes over the role of progesterone (and estrogen) secretion, as the corpus luteum degenerates and disappears from the ovary.

The Testes

The gamete-producing organs of the male are the **testes** which produce **sperm cells** (or **spermatozoa**). In addition to producing sperm, the testes produce and secrete male hormones called **androgens**.

The organization of a testis is illustrated in Figure 13-12. The sperm-producing tissue is arranged to form large numbers of tiny tubules called **seminiferous tubules**. The seminiferous tubules collectively empty into a channel called the **epididymus**. In turn, the epididymus empties into the **ductus deferens** (or **vas deferens**) which conveys sperm out of the testes and into the **urethra** of the **penis** (see also Chapter 14).

Beginning at puberty, sperm cell production in the seminiferous tubules proceeds on a continuous basis. (There is no cycle of sperm production or release comparable to the ovarian cycle of the female.) The production and maturation of each sperm cell, a process called **spermatogenesis**, begins near the wall of the seminiferous tubules and proceeds toward the tubule's lumen (Fig. 13-13). Development begins with cells called **spermatogonia**. These give rise to **primary spermatocytes** which then form **secondary spermatocytes**. In a final stage of maturation, the secondary spermatocytes lose much of their cytoplasm and develop a taillike locomotor appendage called a **flagellum**.

Maturation of sperm is assisted by large cells in the wall of the seminiferous tubules called **Sertoli cells** (Fig. 13-13). These cells supply nutrients to the developing gametes, remove improperly forming sperm cells by phagocytosis, and assist in the withdrawal of cytoplasm from the sperm during the final stages of sperm maturation.

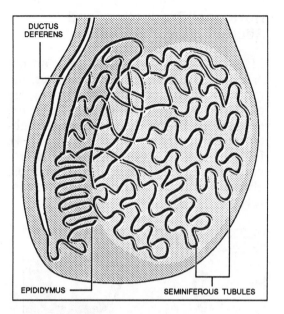

Figure 13-12
Organization of one of the testes. Sperm cells are produced in the seminiferous tubules and move from there into the epididymus, and the ductus deferens.

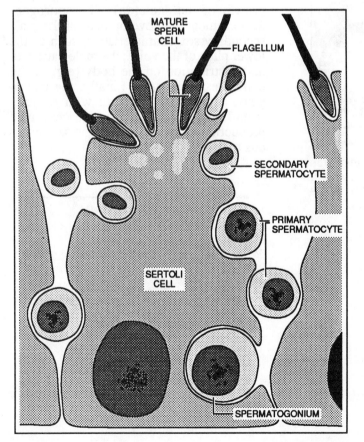

Figure 13-13
The walls of the seminiferous tubules are lined by **Sertoli** cells. These cells assist the development of the sperm. During the last stages of sperm development, the Sertoli cells remove much of the extranuclear cytoplasm from the sperm cells. The tails of mature sperm (top of figure) project into the lumen of the tubule.

Hormone Production and Secretion by the Testes

Maturing sperm and Sertoli cells do not produce or secrete hormones. Rather, the male sex hormones, called *androgens*, are produced by the **cells of Leydig** in the **interstitial tissue** that separates neighboring seminiferous tubules (Fig. 13-14). Like the female sex hormones, the androgens are steroid molecules. The most abundant of the androgens is **testosterone**.

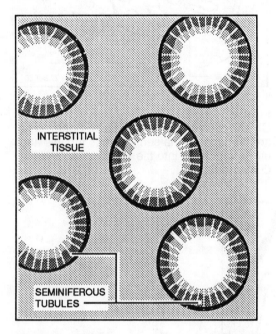

Figure   13-14
Sperm cells are produced in the seminiferous tubules (seen here in cross-section), whereas the male sex hormones are produced by the cells of Leydig in the interstitial tissue.

The production of sperm and the secretion of androgens by the testes are regulated by the same hormonal secretions of the anterior lobe of the pituitary gland that affect the ovarian cycle. Follicle-stimulating hormone promotes sperm production in the seminiferous tubules, whereas luteinizing hormone promotes the secretion of androgens by the cells of Leydig found in the interstitial tissue. Because luteinizing hormone affects the interstitial tissue of the testes, the hormone is also known as **interstitial cell-stimulating hormone (ICSH)**.

Androgens. Testosterone is the most abundant of the androgens, accounting for about 90% of all male sex hormone secreted by the testes. Like the estrogens of the female, the male sex hormones have a diversity of effects. For example, the androgens promote the development of the accessory reproductive organs of the male (i.e., the penis, prostate gland, scrotum, etc; see also Chapter 14). During the teen years, the androgens stimulate muscle development and rapid growth in height (i.e., growth of the skeleton). The androgens also promote the development of the male secondary sex characteristics, such as the lower pitch of the voice, the male pattern of hair distribution on the body (e.g., facial hair, chest hair, etc.). Finally, the androgens, acting together with with follicle-stimulating hormone, promote continued production of sperm cells in the seminiferous tubules.

CHAPTER 14

REPRODUCTION AND INHERITANCE

It has been estimated that the various tissues and organs of an adult human being contain more than 40,000,000,000,000 (that's forty thousand billion!) cells. All of these cells are derived from a single cell called a **zygote** (i.e., the fertilized egg cell) through cell growth and cell division. In many tissues of the body (e.g., the epithelial lining of the digestive organs and the blood-forming tissues), cells continue to grow and divide for most of a person's life. However, in some tissues, such as muscle and nerve, cell division ceases some time after birth, and subsequent tissue growth results from individual cell growth without division.

The ongoing cell growth and division that characterizes most of the body's tissues accounts for the growth of the organism as a whole and the replacement of dying cells. With the exceptions of such tissues as muscle and nerve, replacement of old tissue cells with new ones results in a complete turnover of the body's cellular composition every few years. (A portion of the body's total mass is represented by non-cellular materials that are secreted by cells; for example, most of the mass of bone and carti-

lage is represented by secreted calcium salts and proteins.)

In this chapter, we will explore the processes by which the zygote is formed from sperm and egg cells and the mechanisms by which the cells and tissues of the body are formed. Finally, we will briefly examine some examples of human inheritance that illustrate the processes by which human traits are passed from generation to generation.

FERTILIZATION

Human life begins when an egg cell from the female parent fuses with a sperm cell from the male parent. Fusion of egg and sperm is called **fertilization**, and the new cell that is formed in the process is a human zygote (Fig. 14-1). In order to fertilize an egg, millions of sperm cells must be deposited in the vagina of the female reproductive tract (see later). Because of the high mortality rate of sperm within the female tract, only a few hundred sperm successfully reach the egg cell as it travels from the ovary toward the uterus. Collectively, the

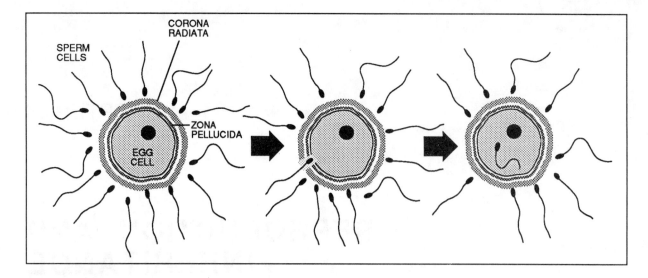

Figure 14-1

Fertilization. Human life begins when an egg cell from the female parent is fertilized by a sperm cell from the male parent. The many sperm that surround the egg cell *(left)* release an enzyme that dissolves the corona radiata. This permits a single sperm cell to penetrate the zona pellucida *(middle)* and enter the egg's cytoplasm *(right)*. The sperm cell nucleus and the nucleus of the egg cell then fuse to form the single nucleus of the zygote.

surviving sperm release an enzyme called **hyaluronidase** that dissolves the **corona radiata** surrounding the egg and allows a single sperm to penetrate the **zona pellucida** and enter the egg's cytoplasm. The sperm cell nucleus and the nucleus of the egg cell then fuse to form the single nucleus of the zygote.

When at fertilization the nuclei of the sperm and egg fuse, the resulting nucleus contains genetic information contributed by each parent. This combination of genetic information will determine the characteristics and properties of the zygote and the billions of cells that are to be derived from the zygote during embryonic and fetal development.

CHROMOSOMES AND GENES

Inheritable information takes the form of individual units called **genes** comprised of the chemical substance **deoxyribonucleic acid** (i.e., DNA; see Chapter 2). It is esti-

mated that the nuclei of human cells contain hundreds of thousands of genes. The genes of a cell nucleus are strung together in linear arrays called **chromosomes**. Chromosomes are not entirely DNA, but also contain large amounts of protein. Although the number of chromosomes in the cell nucleus varies considerably among different animal and plant species, each species has a specific **chromosome number**. In humans, the chromosome number is 46; that is, the hundreds of thousands of genes are apportioned among 46 chromosomes. The chromosome number of a species has no special significance. For example, the number is not necessarily smaller than 46 in organisms simpler than humans (e.g., certain protozoa have over 300 chromosomes); and the number 46 is not reserved for humans (e.g., the potato plant has 48 chromosomes, and so do plum trees and chimpanzees). Within a species, each chromosome exhibits a specific and characteristic shape during the metaphase (see later) of cell division; therefore, each human chromosome can be individually distinguished. For purposes of identification and study, the human chromosomes are assigned specific numbers. (The numbers are 1 through 23, not 1 through 46; the reason for this will be explained later.)

The functional machinery of a cell is its proteins. That is to say, the structure, organization, and functions of cells rest upon the kinds of proteins that are present in the cell. What distinguish one protein from another (chemically and functionally) are the numbers and orders (i.e., sequences) of the various amino acids that comprise the protein's polypeptide chains. It is these two properties of proteins that are encoded in the genes that comprise a chromosome. Thus, for simplicity, a chromosome may be thought of as a sequence of blueprints, each blueprint spelling out the specific structures of cell proteins.

This concept may be illustrated by an example that is already familiar to you. Each of the hemoglobin molecules present in a red blood cell consists (in part) of four polypeptide chains: a pair of *alpha* chains and a pair of *beta* chains (see Fig. 8-3 in Chapter 8). Each of these two types of polypeptide chains is encoded by a gene. The genes are called the *alpha chain gene* and the *beta chain gene*. The alpha chain genes of the cells of the body are found on chromosome number 16, whereas the beta chain genes are found on chromosome number 11. The chromosomal locations of thousands of other human genes are also known.

From the preceding discussion, it is clear that a chromosome may be viewed as a linear sequence of blueprints for cellular proteins. There are, therefore, genes for the alpha and beta chains of hemoglobin, genes for the protein hormone *insulin*, genes for the plasma protein *albumin*, and so on.

Homologous Chromosomes

To further your understanding of the nature of human (and other) chromosomes, let's suppose that the genes that encode insulin, alpha and beta globin chains, albumin, and

a number of other proteins that we'll simply call protein "X," protein "Y," and protein "Z" are all on the same chromosome. (This is not actually true because you already know that alpha globin chain genes are on chromosome number 16 and beta chain genes are found on chromosome number 11; but making such a supposition will help to illustrate several important points.) Such a hypothetical chromosome is illustrated in Figure 14-2.

If, having identified our hypothetical chromosome containing the genes for insulin, alpha and beta globin chains, albumin, and proteins "X," "Y," and "Z," we were now to explore the gene sequences of the other 45 chromosomes in the nucleus, we would find a second chromosome that had the same gene sequence. That is, there are *two* chromosomes in the nucleus containing the genes for insulin, alpha and beta

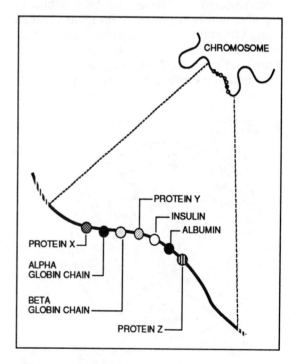

Figure 14-2
A hypothetical human chromosome. A chromosome consists of a series of many genes, each gene encoding the chemical structure of a particular human protein. See text for details.

globin chains, albumin, and proteins "X," "Y," and "Z." The two chromosomes are said to be **homologous** because their gene sequences are the same. In fact, for every chromosome in the nucleus, there is an homologous partner whose sequence of genes (which may number in the thousands) is the same (Fig. 14-3). Therefore, human cells are said to contain 23 pairs of homologous chromosomes. Since the two homologues of a pair are the same, there are only 23 *different* chromosomes in a human cell. (This is why the chromosomes are numbered 1 through 23 and not 1 through 46.) The number of different chromosomes present in a cell is given by the symbol n. Therefore, for humans, $n = 23$.

Because every chromosome has a homologous partner, the genes also occur in pairs. For example, there are two insulin genes—one on each of two homologous chromosomes; there are two albumin genes, two beta globin chain genes, and so on. Organisms (or individual cells) that contain pairs of homologous chromosomes are said to be **diploid**, whereas organisms (or individual cells) that have only one copy

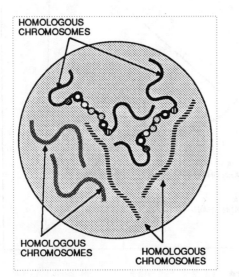

Figure 14-3
A diploid cell nucleus contains pairs of homologous chromosomes. Shown here are three of the 23 pairs of homologous human chromosomes.

of each chromosome (e.g., bacteria, certain other microorganisms, and [as we will see] sperm and egg cells) are said to be **haploid**. Thus, haploid organisms (or individual cells) have n chromosomes, whereas diploid organisms (or individual cells) have $2n$ chromosomes.

Homozygous and Heterozygous Genes

It is usually the case that the two genes occupying equivalent sites on homologous chromosomes are identical; that is, they encode polypeptides with precisely the same number and order of amino acids. Such genes are said to be **homozygous**. However, for a small percentage of gene pairs, there may be small differences between the polypeptide encoded by one gene and the polypeptide encoded by the equivalent gene on the homologous partner chromosome. When these small differences exist, the two genes are said to be **heterozygous**. This notion may be illustrated using an example that is already familiar to you.

In Chapter 9 you learned that there are four human blood types in the ABO series; namely, type A, type B, type AB and type O. These blood types are based upon specific proteins (antigens) that are present in the membranes of red blood cells. A person with type A blood has only the A antigen in the membranes of his red blood cells; a person with type B blood has only the B antigen in the membranes of his red blood cells; a person with type AB blood has both the A antigens and the B antigens in the membranes of his red blood cells; and a person with type O blood has neither A antigens nor the B antigens in the membranes of his red blood cells. The genes that determine ABO blood type occupy equivalent sites on homologous chromosomes. For this gene site, three gene variations occur: (1) gene "A" which encodes antigen A; (2) gene "B" which encodes antigen B; and (3) gene "O" which is not expressed in the form of a detectable antigen. There are, therefore, six different **genotypes** corresponding to the four ABO

214

blood types; these are AA, AO, BB, BO, AB, and OO (see Table 14-1). Every human being has one of these six genotypes, determined at birth when genes contributed by the male and female parents are combined in the zygote.

If a person's genotype is AA (i.e., the person inherited an A gene from each parent), he is said to be *homozygous* and has type A blood. In contrast, a person whose genotype is AO (i.e., the person inherited an A gene from one parent and an O gene from the other parent) is said to be *heterozygous*, although he also has type A blood. During a blood test, both individuals would be found to have type A blood, but genetically one is homozygous and one is heterozygous. Note that all people with type O blood are homozygous and that all people with type AB blood are heterozygous (see Table 14-1). Therefore, because there are three alternate forms of the genes occupying the two ABO sites of homologous chromosomes, there is blood type variation in the human population. Likewise, it is the alternate forms of thousands of other human genes that serve as the source of all variations in the human population. (Another way of saying this is that it is the potentially heterozygous gene pairs that make each one of us different, whereas the genes that are always homozygous are the sources of uniformity among all humans.)

THE CELL CYCLE AND MITOSIS

During fertilization, the sperm and egg cell nuclei fuse, so that chromosomes contributed by the male and female parents are combined in a single nucleus. Almost immediately, the zygote begins the first of many cell divisions that will give rise to the billions of cells of the developing embryo and fetus. That is, the zygote divides to form two cells; each of these cells divides, thereby giving rise to four cells; the four cells become eight cells; the eight become sixteen, and so on. The period of time between the completion of one round of division and the completion of the next round is called a **cell cycle.** As illustrated in Figure 14-4, the cell cycle is divided into two major phases: the **interphase** and the **mitotic phase.** Each of these phases is further subdivided into smaller intervals. In interphase, the chromosomes are diffusely spread through the nucleus and cannot be distinguished as individual bodies. During what is called the **S period** of the interphase, every gene of every chromosome is replicated, so that the genetic complement of the cell is temporarily doubled. In the mitotic phase, the chromosomes condense into visible bodies, and through a process described more fully below, become apportioned between the two daughter cells produced by nuclear and cytoplasmic divi-

TABLE 14-1 HUMAN ABO BLOOD TYPES AND THEIR GENETIC SOURCES

BLOOD TYPE	GENOTYPE	CONDITION
A	AA	Homozygous
A	AO	Heterozygous
B	BB	Homozygous
B	BO	Heterozygous
AB	AB	Heterozygous
O	OO	Homozygous

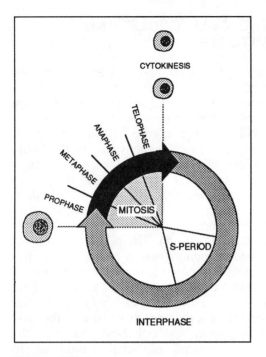

Figure 14-4
The cell cycle. All growing and dividing cells exhibit a cell cycle. The relative lengths of the interphase and mitotic phase vary according to the type of tissue or cell. In this illustration, the length of the mitotic phase is exaggerated for clarity; typically, the mitotic phase accounts for less than 10% of the cell cycle. During the S-period, the nuclear genes (DNA) are replicated. The periods that immediately precede and follow the S-period, called G1 and G2 periods, vary in length. For the series of cell cycles that immediately follow fertilization of an egg cell, the G1 and G2 periods are almost non-existent.

sion. As a result of mitosis, each daughter cell receives one copy of every replicated gene. Therefore, the resulting daughter cells are genetically identical.

For the series of cell divisions that immediately follow fertilization of an egg cell, the duration of the cell cycle is very short. Indeed, the periods that precede and follow the S-period are almost non-existent. However, later in embryonic and fetal development, the length of the cell cycle increases, as more cell growth precedes and

follows the S-period (there is no cell growth during the mitotic phase).

Mitosis

When the zygote divides, the two daughter cells that are formed are genetically identical. When these two cells divide to form four cells, all four progeny are genetically identical. Indeed, virtually all cell divisions produce genetically identical progeny. The mechanism that ensures that identical genetic complements are transferred to the two daughter cells produced by division is a process called **mitosis**.

During mitosis, there is a progressive change in the structure and appearance of the chromosomes. Although mitosis is a continuous process, for convenience it may be divided into four major stages called **prophase, metaphase, anaphase,** and **telophase** (Fig. 14-4). The first stage (i.e., prophase) is characterized by the condensation of the dispersed interphase **chromatin** into visible bodies, namely the chromosomes (Fig. 14-5, stages 1 and 2), the disappearance of the nuclear envelope, and the formation of the **mitotic spindle**. The dramatic nature of the condensation of chromatin into visible chromosomes can be appreciated when one considers that in a human cell, the DNA is 10-15 feet long in its dispersed state but condenses to form chromosomes whose combined length is less than 1/25 of an inch.

The chromosomes are distinguishable by light microscopy and are seen to be composed of two sister **chromatids** held together at the **centromere**. The sister chromatids are the products of the replication of chromosomal DNA during the S-period of the interphase of the cell cycle. Toward the end of prophase, the spindle extends between two poles positioned diametrically opposite one another in the cell. In metaphase, the chromosomes migrate toward the center of the spindle (Fig. 14-5, stage 3). The centromeres are duplicated, so that each chromatid becomes an independent chromosome and is attached to a spindle fiber connected to one of the two poles.

The onset of anaphase is characterized by the movement of the chromosomes toward opposite poles of the mitotic spindle. In the final phase of mitosis, the telophase, the chromosomes reach the poles of the spindle and begin to undergo decondensation. During the telophase, a new nuclear envelope is formed enclosing the chromosomes.

During telophase, a process called **cytokinesis** occurs and divides the cell into two halves, thereby physically separating the two complements of chromosomes. Cytokinesis is a process that is distinct from but synchronized with nuclear division.

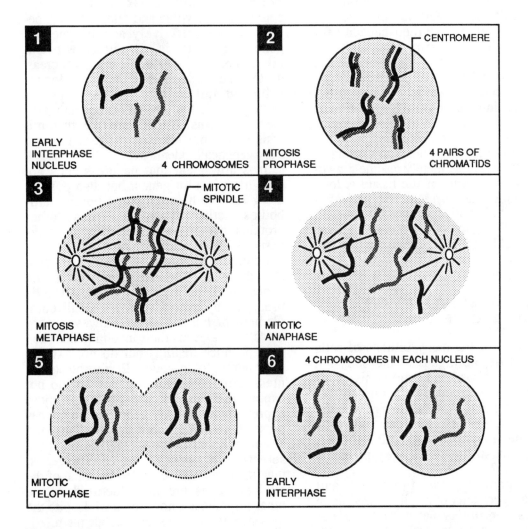

Figure 14-5
Mitosis. Mitosis marks the completion of one cell cycle and the beginning of the interphase of the next cell cycle. Although mitosis is a continuous process, for convenience it may be subdivided into four phases: prophase, metaphase, anaphase, and telophase. Shown here are the fates of 2 of the 23 pairs of homologous chromosomes at specific instances during mitosis. (For simplicity, interphase chromosomes are shown condensed.)

Non-Cycling Cells

Mitosis and cytokinesis produce the billions of cells of the embryo and fetus. Even after birth, these processes continue at a high rate and provide for the growth of the baby and child. As noted earlier in the chapter, the cells of some tissues continue to grow and divide throughout life in order to replace lost and dying cells. For example, skin cells, the epithelial lining of the digestive and respiratory tracts, and blood cells undergo continuous growth and proliferation by mitosis. Other tissues, such as muscle and nerve, lose their capacity for mitosis within a short time after birth and are not replaced if lost by injury or disease. Finally, the cells of some tissues (e.g., liver tissue cells) may exist in a non-growing, non-dividing state for a long period of time. Such cells are said to be temporarily *non-cycling* and have been arrested in the interphase of the cell cycle. Non-cycling or arrested cells may ultimately resume growth and division if some of the tissue is lost. For example, the surgical removal of a portion of the liver is followed by renewed growth and division of the remaining liver tissue, quickly replacing the tissue that had been excised.

TISSUE SPECIALIZATION
(SELECTIVE GENE EXPRESSION)

Virtually all of the many billions of cells that comprise the organs of the body are produced by mitosis and cytokinesis. Since these processes produce cells that are genetically identical, it is fair to ask why the cells in different organs of the body may appear and function differently? Indeed, what is it that distinguishes one type of cell from another?

Fundamentally, the functional and structural specificity of a cell is based upon the types of *proteins* that it produces. For example, erythrocytes are unique in that they are the only cells of the body to produce the protein *hemoglobin*. The islet cells of the pancreas are unique in that they are the only cells of the body to produce the protein *insulin*. B-lymphocytes are unique in that they are the only cells of the body that produce *antibodies*. Only the liver produces and secretes *albumin* into the blood plasma. Only muscle cells produce and organize vast amounts of the proteins *actin* and *myosin* into the thick and thin filaments that form a sarcomere. Each of these specialized proteins (i.e., hemoglobin, insulin, antibodies, albumin, actin, myosin, etc.) is encoded by a pair (or several pairs) of nuclear genes on homologous chromosomes. We might therefore ask whether in the course of embryonic and fetal growth, specific tissues selectively acquire specific gene pairs. That is, as cell growth and division occur, for example in the pancreas, do the islet cells receive the genes for insulin but fail to receive the genes for hemoglobin, albumin, and other non-pancreatic proteins? If this were true, then mitosis would not produce progeny that are truly genetically identical.

It is now clear that pancreas cells not only have insulin genes, but also possess the genes for hemoglobin, albumin, antibodies, actin, myosin, and many other proteins that are not produced by pancreas tissue. Likewise developing blood cells in the bone marrow (e.g., erythroblasts) possess the genes for insulin as well as the genes for hemoglobin. Indeed, all cells produced by mitosis possess a full complement of human genes. However, pancreas cells *express* their insulin genes (i.e., they produce insulin) but do not express their hemoglobin genes. Erythroblasts express their hemoglobin genes but do not express their insulin genes. B-lymphocytes express their antibody genes but do not express their hemoglobin genes. Thus, the key to cell and tissue specialization is the **selective expression** of certain genes; that is, the proteins encoded by some of the nuclear genes are manufactured, while other proteins are not produced despite the presence of the genes that encode them.

MEIOSIS

The two cells produced at the conclusion of

mitosis are genetically identical to one another, and they are also genetically identical to the parental cell from which they were produced. The divisions of the zygote that produce the billions of cells of the embryo are mitotic divisions, so that all progeny have the same 46 (23 pairs of) human chromosomes. Since the zygote is formed by the fusion of a sperm cell from the male parent and an egg cell from the female parent, it is clear that neither the egg nor the sperm that fertilizes it can possess 46 chromosomes. If the egg and sperm each contained 46 chromosomes, then fertilization would produce a zygote with 92 chromosomes; moreover, all of the cells derived by mitosis from this zygote would contain 92 chromosomes. Such a state would imply that the numbers of chromosomes in the nuclei of human cells ought to double with each generation of human beings. Clearly, this is not the case. Rather, the chromosome number is preserved at 46 from one generation to another.

The human chromosome number remains at 46 because sperm and egg cells are **haploid**. That is, these cells contain only 23 chromosomes (half the number of chromosomes found in other cells). The 23 chromosomes present in sperm and egg cells are not a random half of the normal genetic complement of 46 chromosomes. Rather, every egg and sperm cell contains *one member of each homologous pair of chromosomes*. Consequently, when the nuclei of sperm and egg fuse at fertilization, the resulting zygote acquires 23 pairs of homologous chromosomes. Of each homologous pair, one chromosome was derived from the male parent and the other member of the pair was derived from the female parent.

Thus, sperm cells and egg cells are unique in the nature of their genetic complement. These cells are produced only in the reproductive tissues (ovaries of females and testes of males). Whereas all other cells of the body are produced by mitosis, sperm and egg cells are produced by a related process called **meiosis**.

An in-depth discussion of meiosis on a cellular as well as a genetic basis is beyond the scope of this book and is normally treated at length in textbooks of genetics. Therefore, we will limit our discussion of meiosis here to a description of some of the major meiotic events and their implications.

The goal of meiosis is to produce cells (gametes) that are haploid. This is achieved because the paired chromatids of the cell embarking on this process are apportioned among *four* daughter cell nuclei, each nucleus acquiring half the number of chromosomes of the diploid parental cell. Although the resulting cell nuclei contain only half the diploid number of chromosomes, the chromosome set is genetically complete, *because each nucleus acquires one member of each pair of homologous chromosomes*. The homologous chromosomes are assorted randomly at meiotic anaphase (see below), and this accounts in part for the genetic variation that characterizes the human population. Additional genetic variation occurs during the prophase of the first meiotic nuclear division as the result of a phenomenon called "crossing over." The genetic implications of random assortment and crossing over are principal subjects of genetics courses.

Meiosis involves two successive rounds of nuclear division, each round subdivided into its own prophase, metaphase, anaphase, and telophase (Fig. 14-6). In meiotic **prophase I** (Fig. 14-6; stage 2), the chromosomes become visible as condensation of the chromatin begins. As in mitosis, each chromosome can be seen to consist of two chromatids (resulting from replication of all genes during the interphase that precedes the first round of division). In contrast with the events that characterize mitosis, homologous chromosomes become aligned side-by-side so that genes encoding products of similar or identical function are situated adjacent to one another. This phenomenon is called **synapsis**. A this point genes may be exchanged within homologous segments of adjacent chromatids via a mechanism called **crossing over**. When crossing over involves a pair of homozygous genes, no change in the genetic makeup of the chromatids results. However, when crossing

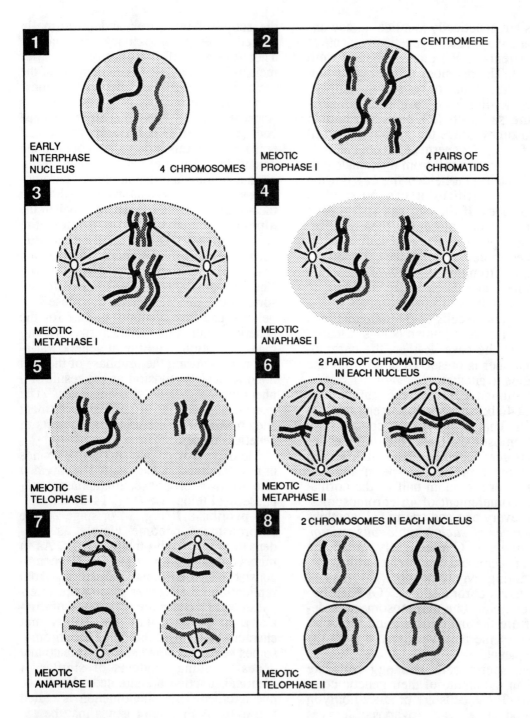

Figure 14-6
Meiosis. During the two rounds of nuclear division that characterize meiosis, a diploid cell gives rise to four haploid cells. Shown here are the fates of 4 (2 homologous pairs) of the 46 chromosomes present in the parental cell. In the first division round, homologous chromosomes are separated and drawn into separate nuclei. In the second division round, paired chromatids are separated and drawn into separate nuclei. (For simplicity, chromosomes are shown condensed in interphase.)

over involves heterozygous gene pairs, the gene sequences of both participating chromatids are altered. Consequently, crossing over creates new combinations of genes in homologous chromosomes.

In meiotic metaphase I (Fig. 14-6; stage 3), the spindle apparatus forms, much as it does in mitosis, and the paired chromatids attach to the spindle fibers arising from opposite poles of the cell.

In meiotic anaphase I (Fig. 14-6; stage 4), homologous chromosomes (but not sister chromatids) separate from each other and move to opposite poles of the spindle. (In mitosis, sister chromatids are separated and drawn to opposite poles.)

Meiotic telophase I (Fig. 14-6; stage 5) brings the first round of meiotic division to a conclusion as the chromosomes aggregate at their respective poles. A new nuclear envelope begins to form around each set of chromosomes and some decondensation of the chromosomes occurs.

The period between the end of telophase I and the onset of meiotic prophase II is usually quite short. In this interphase, the genes (DNA) of the two nuclei produced by the first meiotic division are *not* replicated. The events characterizing meiotic prophase II are similar to mitotic prophase, although each cell nucleus has only half the number of chromosomes as does a cell in mitotic prophase. Each chromosome remains composed of sister chromatids formed in the interphase that preceded meiotic prophase I.

In meiotic metaphase II (Fig. 14-6; stage 6), the paired chromatids migrate to the center of the newly-forming spindle. Then, in meiotic anaphase II (Fig. 14-6; stage 7), sister chromatids separate from one another and are drawn to opposite poles. Finally, in meiotic telophase II, the separated chromosome groups are enclosed in a newly developing nuclear envelope (Fig. 14-6; stage 8) and begin decondensation.

SPERMATOGENESIS AND OOGENESIS

In both males and females, gene replication followed by meiosis converts a single diploid cell into four haploid cells. In males, the process occurs in the testes and is called **spermatogenesis**; in females, the process occurs in the ovaries and is called **oogenesis**.

Spermatogenesis

Spermatogenesis occurs in the **seminiferous tubules** of the testes by the meiotic division of cells that form the tubules' walls. These cells are called **spermatogonia** (Fig. 14-7, but also see Figs. 13-12 and 13-13 in Chapter 13). Prior to meiotic prophase I, the chromosomes of the spermatogonia are replicated, thereby forming what are called **primary spermatocytes**. Meiotic division I converts each primary spermatocyte into two **secondary spermatocytes**. Meiotic division II converts each secondary spermatocyte to two haploid **spermatids**. Therefore, four spermatids are produced by the meiotic divisions of one spermatogonium. Each spermatid undergoes a series of morphological changes in which most of the cytoplasm is lost and the cell develops a tail-like **flagellum** that will act to propel the cell once it is deposited in the female reproductive tract. The mature gamete is called a **spermatozoan** or **sperm cell**. The meiotic changes of the spermatogonium and the maturation of spermatids into sperm cells are assisted by the actions of other cells in the walls of the seminiferous tubules called **Sertoli cells** (see Fig. 13-13 in Chapter 13).

All four sperm cells produced by meiotic division are **viable**; that is to say all are potentially capable of fertilizing an egg cell. In contrast (see later), meiosis in the female produces a single viable gamete (the egg cell) and three non-viable (non-functional) cells called **polar bodies**.

During the first meiotic division of spermatogenesis, genes present in one of the two sister chromatids comprising each chromosome are exchanged with corresponding genes in one of the sister chromatids comprising the homologous chromosome (i.e., "crossing over" occurs).

221

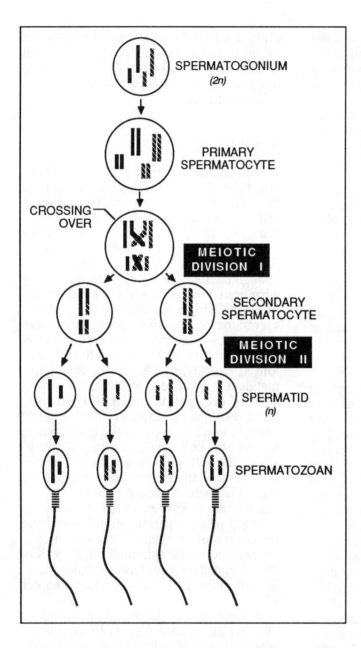

Figure 14-7
Spematogenesis. During spermatogenesis, meiosis produces 4
haploid, viable sperm cells from a
single diploid spermatogonium. For
simplicity, only two of the 23 pairs of
homologous human chromosomes
are shown here. Notice that during
meiotic division I, crossing over
results in an exchange of homologous chromosomal segments.
Although only one crossing over
point is shown here, there may be
many.

Exchanges involving heterozygous gene
pairs are one of the sources of genetic variety among the spermatids that are formed.
Additional genetic variation results from
random assortment of chromosomes among
the daughter nuclei. Thus, the four sperm
cells produced by the meiosis of one spermatogonium are genetically different.

In males the conversion of spermatogonia into primary spermatocytes begins as
early as during embryonic development.
However, meiosis proceeds no further until
puberty. The initial stages of spermatogenesis appear to be promoted by the hormone
testosterone, whereas later stages of spermatogenesis are promoted by follicle-stimulating hormone (again, see Chapter 13).
Once spermatogenesis begins in the male at
puberty, it takes place continuously for 50
or more years.

Oogenesis

Oogenesis, the production by meiosis of haploid egg cells or **ova** in the ovary, is depicted in Figure 14-8. Although the process is fundamentally similar to spermatogenesis, there are some important differences. Oogenesis begins with diploid **oogonia** that are first converted into **primary oocytes**. The primary oocytes then begin the first meiotic division. This division is said to be "unequal" in that one of the daughter cells receives nearly all of the cytoplasm of the parental cell, while the other daughter cell receives very little cytoplasm (Fig. 14-8). The cell that receives the bulk of the cytoplasm is called a **secondary oocyte** and will give rise at the completion of the second meiotic division to a functional egg cell. The other cell, called a **first polar body**, will not produce a viable ovum.

Both the primary oocyte and the first polar body undergo the second meiotic division (Fig. 14-8). The two haploid products of the first polar body are also called polar bodies and likewise are non-functional. Division of the primary oocyte is also unequal and produces a **secondary oocyte** and a non-functional **second polar body**. The secondary oocyte, which is haploid, has the capacity to mature into a viable (potentially fertilizable) egg cell. Therefore, whereas meiosis in males produces four viable gametes, meiosis in females produces one viable gamete (and three functionless polar bodies).

During the first meiotic division of oogenesis, genes present in one of the two sister chromatids comprising each chromosome are exchanged with corresponding genes in one of the sister chromatids comprising the homologous chromosome (i.e., "crossing over" occurs; Fig. 14-8). Pairs of sister chromatids and whole chromosomes are also randomly assorted among the two progeny of each division. Thus, like sperm, the egg cell and three polar bodies are genetically different.

Oogenesis gets underway during the embryonic development of a female, so that at birth the ovaries of the average female contain hundreds of thousands of primary oocytes. Each primary oocyte is enclosed by a simple layer of supportive cells, thereby forming a **primary follicle** (see Chapter 13). No further development of follicles or egg cells occurs until puberty.

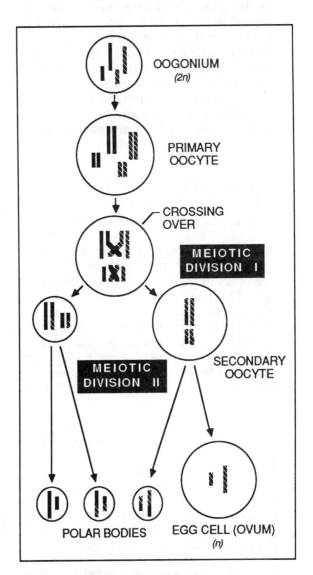

Figure 14-8
Oogenesis. During oogenesis, meiosis produces 1 haploid, viable egg cell and three functionless polar bodies from a single diploid oogonium. For simplicity, only two of the 23 pairs of homologous human chromosomes are shown here. During meiotic division I, crossing over results in an exchange of homologous chromosomal segments.

223

At puberty, under the influence of pituitary hormones, the female begins her succession of ovarian (or menstrual) cycles. During each ovarian cycle, meiosis is resumed by the prospective ovum of the **Graafian follicle** (see Chapter 13), and by the time of ovulation meiosis has proceeded to the secondary oocyte stage. Following ovulation, meiosis progresses to meiotic metaphase II but is arrested at that stage. The second round of meiotic division and cytokinesis are completed only if the prospective ovum is fertilized by a sperm cell. If the prospective egg cell is not fertilized, it is absorbed and digested by the lining of the oviduct (see below).

ORGANIZATION OF THE FEMALE AND MALE REPRODUCTIVE TRACTS

In females, the organs of the reproductive system remain in an infantile condition until the onset of puberty, which occurs around age 13. The anatomical and chemical changes that occur during puberty are believed to be initiated by events taking place in the brain, although the precise nature of these events remains obscure. It is apparent that well before the onset of puberty, the ovaries are already in a state capable of re-leasing egg cells. This has been amply demonstrated through experiments in which injections of pituitary gonadotropic hormones into immature laboratory animals trigger the early onset of puberty, the initiation of the ovarian cycle, and the secretion of the ovarian hormones. Whereas, the pituitary glands of immature animals contain these gonadotropins, under normal conditions the release of these hormones is somehow deferred.

In females, the succession of ovarian cycles continues on a regular basis until the age of 45 to 55 years, when the cycle starts to become irregular and eventually ceases altogether. This is known as the **menopause** and may extend over a period of several years. (In males, there is no dramatic change in reproductive function; instead, testicular function declines very slowly.) During the menopause, sensitivity of the ovaries to pituitary gonadotropins disappears, and there is no further development of follicles. The halt to follicle development is accompanied by a strong reduction in the secretion of estrogen.

The organization of the reproductive tract of an adult female is depicted in Figure 14-9 (see also Fig. 14-11). The major organs of the tract are the **vagina**, **uterus**, **Fallopian tubes** (or **oviducts**) and the **ovaries**. The vagina is a passageway that

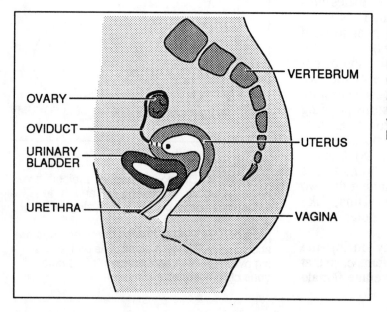

Figure 14-9
Side view of the organization of the female reproductive tract. Associated with each ovary (only the right ovary is depicted here) is a Fallopian tube, which conducts ova (egg cells) toward the uterus. The vagina leads from the uterus to the body surface.

VERTEBRUM

OVARY

OVIDUCT

URINARY BLADDER

URETHRA

UTERUS

VAGINA

extends upwardly from the body surface into the thick-walled and muscular uterus. At its junction with the uterus, the vagina narrows to form the **cervix**. Entering the uterus on either side from above are the Fallopian tubes. Each of these narrow channels begins near the outer edge of an ovary, curves over the ovary's upper surface and then descends toward the uterus.

The vagina acts to receive sperm from the male during **copulation** (or **coitus**). From the vagina, the sperm make their way through the uterus and into the Fallopian tubes. If the tubes contain an egg cell recently released from the ovary (see below), and if the sperm successfully fertilize the egg, the resulting zygote is swept toward the uterus and implanted in the uterine wall. The growth and development of the embryo and fetus take place within the uterus.

The organization of the male reproductive system is shown in Figure 14-10. Sperm are carried out of each of the testes by narrow tubes called **vas deferens**. Just below and behind the urinary bladder, each vas deferens forms an **ejaculatory duct**. Located in this region is the **seminal vesicle** which adds fluid to the sperm. Beyond the seminal vesicle, the ejaculatory ducts lead into the urethra near the urethra's exit from the urinary bladder. In this region, the **prostate** and **bulbourethral**

glands add additional fluid to the semen. The urethra extends into and conveys the sperm through the **penis**.

FERTILIZATION AND IMPLANTATION

The egg cell that is released from the ovary on day 14 of the ovarian cycle is drawn by fluid currents into a Fallopian tube. These currents are created by the beating of cilia that cover the finger-like projections (**fimbriae**) that exist at the mouth of the Fallopian tube (Fig. 14-11). The egg cell then begins a 3-day journey along the Fallopian tube toward the uterus. Since the egg cell has no independent means of locomotion, it relies on the beating cilia of the cells that line the oviduct in order to be moved along.

Following ovulation, the egg cell remains viable for 12 to 24 hours. Since the journey to the uterus requires about three days, this implies that the egg must be fertilized somewhere between the mouth of the oviduct and approximately the first one-third of the oviduct's length (see Fig. 14-11). If the egg cell is not fertilized, it is absorbed and destroyed by the cells lining the second half of the oviduct.

Figure 14-10
Side view of the organization of the male reproductive tract. Sperm produced in the testes (only the right testis is shown here) are conveyed into the urethra by the vas deferens. To the sperm are added the secretions of the seminal vesicle, prostate gland, and bulbourethral gland .

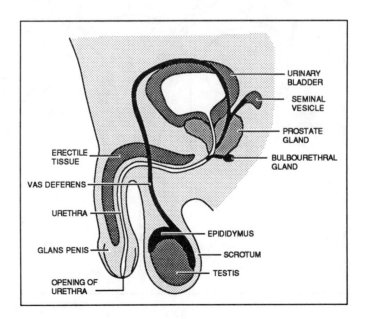

When sperm are deposited in the vagina during copulation, they propel themselves into the uterus and into the Fallopian tubes. Because sperm cells are motile and can move relatively quickly, they can reach the mouth of each of the Fallopian tubes within a few hours. After being deposited in the female reproductive tract, sperm remain viable for about three days. This implies that pregnancy is possible during an interval of about four days during each ovarian cycle. In the average female, this "fertile interval" extends from about day 12 until day 15. For example, if sperm are deposited in the female reproductive tract on day 12, they will quickly reach the ends of the oviducts, where they will remain viable (capable of fertilizing an egg cell) until ovulation occurs on day 14. In such an instance, the egg cell will be fertilized almost immediately following ovulation. On the other hand, if sperm enter the female reproductive tract on day 15, they may reach and fertilize the egg cell while it is in the lingering hours of its viability.

During fertilization, the sperm cell penetrates the corona radiata and zona pellucida of the egg and enters the egg cell's cytoplasm (see Fig. 14-1). The sperm cell nucleus and egg cell nucleus then fuse to form a single nucleus. Since the nuclei of the two gametes are haploid, their fusion creates a diploid nucleus. The 23 chromosomes donated by the sperm represent one of each of the 23 homologous pairs originally present in the spermatogonium. By the same token, the 23 chromosomes donated by the egg cell nucleus represent one of each of the 23 homologous pairs originally present in the oogonium. Therefore, the fusion of the sperm and egg cell nuclei at fertilization produces a **zygote** whose nucleus contains 23 pairs of homologous chromosomes.

Immediately after the two nuclei have fused, the chromosomes are replicated and the first mitotic division is initiated. This division produces two diploid cells (Figure 14-12). A second mitotic division quickly follows, thereby producing four diploid cells. This process continues as the fertilized egg slowly proceeds toward the uterus.

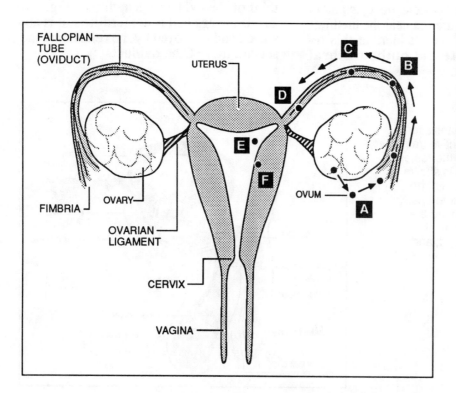

Figure 14-11
Front view of the female reproductive tract. A, B, C, D, E, and F show the egg cell (zygote) as it progresses toward (and is eventually implanted in) the uterine wall (see also Figure 14-12).

Because the dividing zygote does not grow between successive divisions, the cells produced by these rounds of mitosis are consecutively smaller and smaller. (Growth in size does not begin until implantation in the wall of the uterus, where the rich supply of blood provides the raw materials needed for growth.) Continued mitosis produces a ball of cells, called a **morula**, that is no greater in size than the original zygote (Fig. 14-12). By the time the uterus is reached, the morula is transformed through rearrangements of the positions of the cells into a **blastocyst**. The organization of a blastocyst is seen in Figure 14-12F. As seen in the figure, the blastocyst consists of an **inner cell mass** facing a fluid-filled, enclosed chamber (i.e., the **blastocoel**). The blastocyst re-

mains in this state within the uterus for several days before **implantation** occurs. On about the seventh day following ovulation (corresponding to day 21 of the ovarian cycle), the blastocyst attaches to the wall of the uterus. Digestive enzymes secreted by cells forming the outer layer of the blastocyst allow the blastocyst to burrow more deeply into the uterine wall, where nutrients are more readily acquired. The embryo that now begins to form arises from the cells of the inner cell mass.

TWINS

About one in every one hundred pregnancies results in the birth of twins. There are two types of twins: **fraternal twins** and

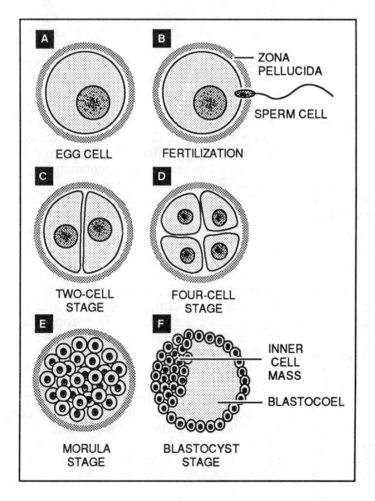

Figure 14-12
Developmental stages that immediately follow ovulation. The six stages shown here (i.e., A through F) correspond to the six lettered positions of the ovum prior to and following fertilization shown in Figure 14-11.

identical twins. Fraternal twins result from the rare occurrence in which two eggs are released from the ovaries during a single ovarian cycle and both are fertilized by sperm cells. Since two zygotes are formed, fraternal twins are also known as **dizygotic twins**. Each zygote is transformed into a blastocyst, and the two blastocysts independently implant in the wall of the uterus. Because they are formed from separate zygotes, fraternal twins are genetically no more alike than are any other family siblings. They may or may not be of the same sex.

Identical twins are genetically identical (which of course implies that they are the same sex). Such twins arise when the inner cell mass of the blastocyst separates into two separate clusters of cells. Each cluster develops into a complete embryo and fetus. Because identical twins develop from a single zygote, they are also called **monozygotic twins**.

SOME FUNDAMENTAL PRINCIPLES OF INHERITANCE

Inheritance of ABO Blood Type and Rhesus Factor

ABO Series Blood Type. As noted earlier in this chapter and also in Chapter 9, there are four different blood types corresponding to the ABO antigen series. These are blood types A, B, AB, and O. Which one of these blood types a given individual has is determined genetically and is called the **phenotype**. A person's phenotype is readily identified by a clinical blood test. We also saw earlier in this chapter that there are three different forms of the gene that encodes an ABO series antigen: gene A encodes antigen A; gene B encodes antigen B; and gene O encodes a clinically undetectable product. Because there is more than one form of the gene, the ABO series genes belong in the class of potentially heterozygous genes. Because there are three versions of the gene, there are six possible **genotype** variations corresponding to the four phenotypes (see

Table 14-1). That is, a person whose is phenotypically blood type A may be genotypically AA (homozygous) or genotypically AO (heterozygous). A person whose is phenotypically blood type B may be genotypically BB (homozygous) or genotypically BO (heterozygous). All persons with blood type AB are heterozygous, and all persons with blood type O are homozygous (i.e., they are OO).

Now, consider a male who has type AB blood. In all of his diploid cells (including the spermatogonia in his testes), the homologous pair of chromosomes carrying the ABO series genes would reveal the A gene on one member of the pair and the B gene on the other member. When sperm cells are produced by spermatogenesis, only one member of each homologous pair of chromosomes ends up in each sperm cell. Therefore, half of the sperm that he produces will have the A-gene containing homologue and the other half of his sperm will have the B-gene containing homologue. Consequently, depending upon which of the two types of sperm fertilizes an egg cell, he will either pass on the A-gene to the zygote or will pass on the B-gene. The same rules apply to a female who has type AB blood. Oogenesis will produce two types of egg cells with respect to the ABO series genes: A-gene containing egg cells and B-gene containing egg cells.

To illustrate the fundamental rules of inheritance, let's explore the various ABO blood types that are possible in the children born to parents who have type AB blood. The chances that a zygote is formed from an A-gene containing sperm are 50:50 (i.e., the odds are 0.5), and the chances that the zygote is formed from an A-gene containing egg are also 50:50. Therefore, the chances of the zygote getting an A-gene from both parents are 1:4 (i.e., $0.5 \times 0.5 = 0.25$). The same odds favor a zygote with B-genes from both parents. However, the odds of a zygote getting an A-gene from one parent and a B-gene from the other parent are 1 in 2 (i.e., 50:50). This is because there are two different ways in which this can occur; namely, if the zygote receives an A-gene from the sperm and a B-gene from the egg, or if the zygote receives a B-gene from the sperm and an A-gene from the egg. Thus, from a statistical point of view, one-half of the children born to type AB parents will have AB blood, one-fourth will have type A blood, and the remaining one-fourth will have type B blood.

All of these conclusions are more easily drawn and understood, if we use what is called a **Punnett square** to explore the potential genotypes of the zygotes formed. Such a Punnett square is shown in Figure 14-13. The male phenotype and genotype are shown to the left of the square and in the neighboring subsquares are placed the letters corresponding to the two types of sperm cells that are produced. Above the Punnett square are identified the phenotype and genotype of the female parent. In the subsquares below are placed the letters corresponding to the two types of egg cells that are produced. Finally, in each of the remaining four squares are placed two letters corresponding to the gene donated by the egg (above) and the gene donated by the sperm (to the left).

Figure 14-14 is a Punnett square showing all of the possibilities that exist among the offspring of an AB father and a type O mother. Notice that in this case, none of the zygotes (and therefore none of the children) will have the same blood type as their parents.

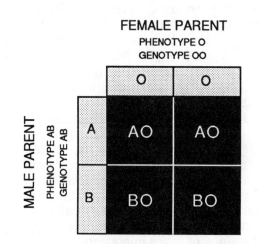

Figure 14-14
A Punnett square showing the genotypes possible among the zygotes formed by parents with type AB and type O blood.

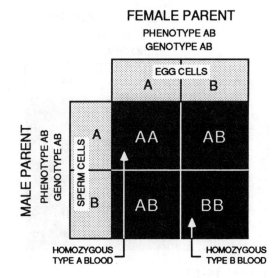

Figure 14-13
A **Punnett square** showing the genotypes possible among the zygotes formed by parents with type AB blood.

Rhesus Blood Type (Rhesus Factor). The Rhesus factor is another example of an inherited cell membrane antigen. The Rhesus antigen is named after the the *Rhesus* monkey (in which the antigen was first discovered). Every person is either *Rhesus positive* (i.e., **Rh⁺**) or *Rhesus negative* (i.e., **rh⁻**). Two genes occupying homologous sites of homologous chromosomes are involved in determining a person's Rhesus type. One of these genes is expressed through the production of the Rhesus antigen (i.e., the Rh⁺ gene); the other gene (i.e., the rh⁻ gene) is not expressed as a detectable antigen. Therefore, with regard to this pair of genes, there are three possible genotypes (see Fig. 13 in Chapter 9); namely, Rh⁺Rh⁺ (an

Rh+ gene inherited from each parent); Rh+rh- (an Rh+ gene inherited from one parent and an rh- gene inherited from the other), and rh-rh- (rh- genes inherited from both parents). So long as at least one Rh+ gene is inherited, the person's blood cells will contain the Rhesus antigen and be classified as Rhesus positive (about 85% of the population). If no Rh+ gene is present, the person is Rhesus negative (about 15% of the population).

As discussed in Chapter 9, the Rhesus factor takes on special significance during pregnancies in which the mother is rh- and the developing fetus is Rh+. Under these circumstances, the mother may become sensitized to Rhesus antigens that enter her bloodstream from the fetus. The mother's immune response results in the production of antibodies against Rhesus antigens. These antibodies may lead to the destruction of fetal blood cells either in the current pregnancy or in a similar ensuing pregnancy. The damage done to the fetal circulation, called *Erythroblastosis fetalis*, can be fatal. Knowing one's Rhesus blood type is, therefore, of major importance.

To illustrate the inheritance of the Rhesus factor, let's consider the various possibilities among the children of a female who is rh- and a male who is Rh+. Because the female is rh-, we automatically know that she is homozygous (i.e., rh-rh-). However, the Rh+ male may be either homozygous (i.e., Rh+Rh+) or heterozygous (i.e., Rh+rh-). For purposes of this illustration, we'll suppose that he is homozygous. The Punnett square of Figure 14-15 shows the results. Because all of the eggs produced by the female will contain the rh- gene, and because all of the sperm produced by this male will contain the Rh+ gene, this means that the genotypes of all of the children will be Rh+rh-. Therefore, all of the children will have Rhesus positive blood and each pregnancy must be monitored for the possibility of Erythroblastosis fetalis.

As shown in Figure 14-16, if the male parent were heterozygous, then (on a statistical basis) only half of the children would have Rhesus positive blood (the other half would have Rhesus negative blood).

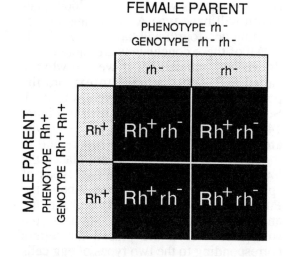

Figure 14-15
A Punnett square showing the genotypes possible among the zygotes formed by parents with rhesus positive (father) and Rhesus negative (mother) blood.

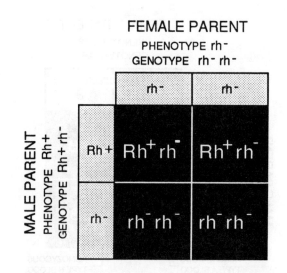

Figure 14-16
A Punnett square showing the genotypes possible among the zygotes formed by parents with rhesus positive (father) and Rhesus negative (mother) blood. In this example, the father is heterozygous

DETERMINATION OF SEX, AND SEX-LINKED INHERITANCE

Sex Determination

The two chromosomes that constitute an homologous pair are readily identified within the metaphase nucleus of a cell, because these chromosomes are the same size and shape. However, in the diploid cells of males, the homologous partners of only 22 of the 23 different chromosomes can be identified on the basis of their similar shape and size. That is, there are two chromosomes left over after you have matched up the other 22 pairs of homologues on the basis of their physical appearances. The two odd chromosomes are the **sex chromosomes**; that is, they are the chromosomes containing genes that determine the individual's sex. The other 22 pairs of chromosomes are called **autosomes** (or autosomal chromosomes).

In males, one of the two sex chromosomes is much larger than the other and contains many more genes. This chromosome is called an **X-chromosome**. The other, smaller sex chromosome is called a **Y-chromosome**. Thus, the 46 chromosomes present in the diploid tissue cells of males consist of 22 pairs of autosomal chromosomes, one X-chromosome, and one Y-chromosome.

In females, the situation is different. In the diploid tissue cells of females, there are 22 pairs of autosomal chromosomes and *two* X-chromosomes. That is, female cells do not contain a Y-chromosome; rather, they contain two X-chromosomes. With regard to their sex chromosomes, males may be described as **XY**, whereas females are **XX**.

When spermatogenesis in males produces haploid sperm, each sperm cell receives either an X-chromosome or a Y-chromosome from the diploid spermatogonium. In females, oogenesis produces haploid egg cells containing one of the two X-chromosomes originally present in the oogonium. The Punnett square shown in Figure 14-17 shows that the sex of the zygote formed when an egg cell is fertilized by a sperm cell depends upon whether the sperm contains a Y-chromosome or an X-chromosome. If the sperm contains a Y-chromosome, then the zygote is male. This is because the zygote contains one X-chromosome (donated by the egg cell) and one Y-chromosome (donated by the sperm cell). If the sperm contains an X-chromosome, then the zygote is female (because the zygote contains two X-chromosomes, one donated by the egg cell and one donated by the sperm cell). Since the sex of the zygote depends upon which sex-chromosome is present in the sperm cell, it is said that "the father determines the sex of the child."

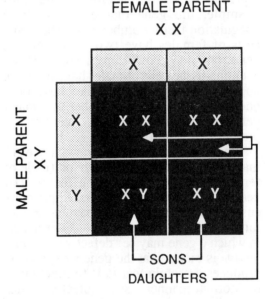

Figure 14-17
Punnett square depicting the determination of the sex of the zygote. In this instance, "X" and "Y" represent whole chromosomes (i.e., the sex chromosomes), not individual genes. Note that from a statistical point of view, half of the zygotes will be male (YX) and half will be female (XX).

Sex-Linked Inheritance

As already noted, the sex chromosomes contain genes that determine an individual's

231

sex. X-chromosomes, however, also contain genes that encode characteristics unrelated to sex; Y-chromosomes lack these genes (in fact, very few genes have been identified on Y-chromosomes). For example, the genes that encode blood coagulation factors VIII (antihemophilic factor) and IX (Christmas factor) are located on the X-chromosome (see also Chapter 8). Genes that are present on X-chromosomes but absent from Y-chromosomes are called **sex-linked genes**. As we will see, these genes give rise to the peculiar patterns of **sex-linked inheritance**.

The fact that certain genes are present on X-chromosomes but absent from Y-chromosomes creates certain peculiarities with regard to inheritance patterns. Consider, for example, the gene that encodes blood coagulation factor number VIII. The tissue cells of females have two copies of this gene, one on each X-chromosome. However, the tissue cells of males have only one copy of the gene, because they have only one X-chromosome (remember that many genes found on X-chromosomes are not present on Y-chromosomes).

From a practical point of view, it doesn't make much difference if a person (male or female) has one copy of the factor VIII gene or two copies of the gene. Either copy ensures that normal blood coagulation factor is produced. However, consider a case in which a gene may be "defective" (by *defective* is meant that the gene's expression produces a protein that fails to function in the normal manner). Such defective genes arise through mutations that occur either spontaneously (i.e., naturally) or which are induced by chemical or physical forces (such as exposure to mutagenic chemicals, radiation, etc.). In males, a single defective sex-linked gene implies that there will be no correct product of that gene's expression. In contrast, in females, when the sex-linked gene of one X-chromosome is defective, the normal gene present on the other X-chromosome ensures that the correct protein encoded by the gene is produced. The male would exhibit the symptoms of the presence of the defective sex-linked gene, whereas the female would not.

Abnormalities that are associated with sex-linked genes are called **sex-linked diseases**. Hemophilia A (failure to produce adequate amounts of coagulation factor VIII) and hemophilia B (failure to produce adequate amounts of coagulation factor IX; again, see Chapter 8) are examples of sex-linked diseases. These diseases are much more common in males than in females because they can stem from the presence of a single defective gene. The only way that a female exhibits a sex-linked disease (for example, hemophilia A) is if the genes on both X-chromosomes are defective.

The Punnett squares of Figures 14-18, 14-19, and 14-20 illustrate patterns of inheritance of hemophilia. In these figures, the normal coagulation factor gene is represented by an upper case "H" and the defective gene by a lower case "h." The letter "Y" is used to show Y-chromosomes, which do not carry either gene.

Figure 14-18 shows the progeny of a male who has hemophilia and a perfectly normal female. Note that the contents of

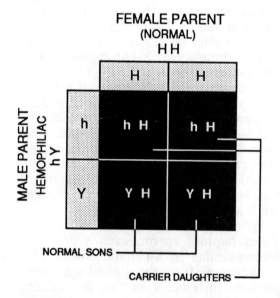

Figure 14-18
Distributions of the normal and abnormal blood coagulation genes among the children of a hemophiliac male and normal female.

each sub-square reveal both the blood co-agulation characteristics of the offspring and also the offspring's sex. Because they inherit the Y-chromosome from their father, all sons will have blood that coagulates normally. In contrast, all daughters will inherit from their father an X-chromosome that contains the defective blood coagulation gene. Since these daughters will also inherit a normal X-chromosome from their mother, their blood will coagulate properly. These daughters are called "carriers" because they carry the defective sex-linked gene on one of their X-chromosomes and may pass it on to one of their children.

The Punnett square of Figure 14-19 illustrates the possible progeny of a normal male and a female who is a carrier of the hemophilia gene. Note that on a statistical basis, half of the sons will be hemophiliacs and half of the daughters will be carriers.

In neither of the two previous illustrations does the possibility exist that a daughter will be a hemophiliac. Indeed, sex-linked diseases in females are very rare. In order to suffer from a sex-linked disease, a female must inherit a defective gene-containing X-chromosome from her mother and a defective gene-containing X-chromosome from her father. With regard to hemophilia, this can occur only if (1) the father is a hemophiliac and the mother is a carrier, or (2) both the father and mother are hemophiliacs. The former possibility is illustrated in the Punnett square of Figure 14-20.

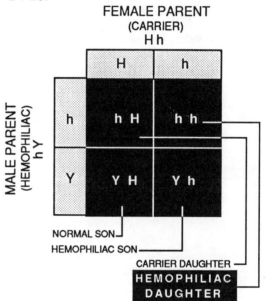

Figure 14-20
Distributions of the normal and abnormal blood coagulation genes among the children of a hemophiliac male and a female who is carrying a defective sex-linked blood coagulation gene.

Figures 14-18, 14-19, and 14-20 show the peculiar patterns of inheritance that characterize sex-linked genes. In addition to hemophilia, several other human diseases are believed to result from sex-linkage. These include (1) defective green color vision (one of several forms of color-blindness), (2) juvenile muscular dystrophy, and (3) Hunter's syndrome (characterized by mental retardation and abnormal physical development).

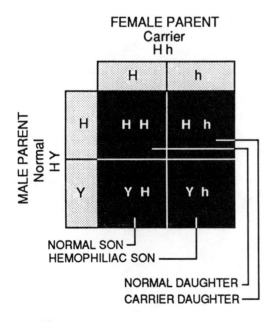

Figure 14-19
Distributions of the normal and abnormal blood coagulation genes among the children of a normal male and a female who is carrying a defective sex-linked blood coagulation gene.

INDEX